Understanding
Correlation Matrices

Quantitative Applications in the Social Sciences

A SAGE PUBLICATIONS SERIES

Understanding Correlation Matrices

Alexandria Hadd

Spelman College

Joseph Lee Rodgers

Vanderbilt University

Quantitative Applications in the Social Sciences, Volume 186

Los Angeles | London | New Delhi
Singapore | Washington DC | Melbourne

FOR INFORMATION:

SAGE Publications, Inc.
2455 Teller Road
Thousand Oaks, California 91320
E-mail: order@sagepub.com

SAGE Publications Ltd.
1 Oliver's Yard
55 City Road
London EC1Y 1SP
United Kingdom

SAGE Publications India Pvt. Ltd.
B 1/I 1 Mohan Cooperative Industrial Area
Mathura Road, New Delhi 110 044
India

SAGE Publications Asia-Pacific Pte. Ltd.
18 Cross Street #10-10/11/12
China Square Central
Singapore 048423

Printed in the United States of America

Library of Congress Cataloging-in-Publication Data

Names: Hadd, Alexandria, author. | Rodgers, Joseph Lee, 1953- author.

Title: Understanding correlation matrices / Alexandria Hadd, Spelman College, Joseph Lee Rodgers, Vanderbilt University.

Description: First Edition. | Thousand Oaks : SAGE Publications, Inc, 2020. | Includes bibliographical references.

Identifiers: LCCN 2020031187 | ISBN 9781544341095 (paperback) | ISBN 9781544341071 (epub) | ISBN 9781544341088 (epub) | ISBN 9781544341101 (ebook)

Subjects: LCSH: Correlation (Statistics) | Matrices.

Classification: LCC HA31.3 .H333 2020 | DDC 519.5/37—dc23

LC record available at https://lccn.loc.gov/2020031187

Acquisitions Editor: Helen Salmon
Editorial Assistant: Elizabeth Cruz
Production Editor: Natasha Tiwari
Copy Editor: QuADS Prepress Pvt. Ltd.
Typesetter: Hurix Digital
Proofreader: Liann Lech
Indexer: Integra
Cover Designer: Candice Harman
Marketing Manager: Victoria Velasquez

20 21 22 23 24 10 9 8 7 6 5 4 3 2 1

TABLE OF CONTENTS

SERIES EDITOR'S INTRODUCTION

In 1988, Joseph Rodgers and Alan Nicewander published an article in *The American Statistician* that described 13 different ways to look at the correlation coefficient. This classic article is one of my favorites. It holds up one of the workhorses of modern statistics and examines it from multiple perspectives, twisting and turning it to develop new understandings and insights. In *Understanding Correlation Matrices*, Alexandria Hadd and Joseph Rodgers team up to provide a parallel treatment for the correlation matrix.

Correlation matrices (along with their unstandardized counterparts, covariance matrices) underlie much of the statistical machinery in common use today. Multiple regression models, confirmatory and exploratory factor analysis, principal components analysis, and structural equation models often start with a correlation (or covariance) matrix. The correlation matrix is so common, we hardly ever give it a second look unless, of course, there are problems that prevent us from estimating the model we have specified. This book shines a light on the correlation matrix, examining it from multiple perspectives: mathematical, statistical, and geometric.

A correlation matrix is more than a matrix filled with correlation coefficients. The value of one coefficient in the matrix puts constraints on the values of the others. This is a major theme of the book. As Hadd and Rodgers explain, a "true" correlation matrix must be symmetric, with diagonal elements equal to one, off-diagonals containing correlation coefficients bounded by -1 and 1, and nonnegative eigenvalues. If the first three conditions are met, but not the fourth, the matrix is a pseudo-correlation matrix. This can happen, for example, when the correlation coefficients are estimated based on different samples. Hadd and Rodgers explore these and other characteristics of correlation matrices in chapters devoted to statistical hypothesis testing on correlation matrices, methods for graphing correlation matrices, and the geometry of correlation space.

Although some of the topics are advanced, the book is written to be accessible to readers with no background in linear algebra. The key points are illustrated with a wide range of lively examples, including correlations between intelligence measured at different ages through adolescence;

correlations between public health expenditures, health life expectancy, adult mortality, and other country characteristics; correlations between well-being and state-level vital statistics; correlations between the racial composition of cities and professional sports teams; and correlations between childbearing intentions and childbearing outcomes over the reproductive life course.

One of the reasons that I so thoroughly enjoy methods and statistics is that topics can be understood on many levels. The deeper you dig, the more you learn. This book is a great illustration: Even sophisticated readers will find a new appreciation for an old friend, the correlation matrix. I invite you to join Ali and Joe, and dig deep.

Barbara Entwisle
—Series Editor

PREFACE

This book was written to present, for the first time, a description of the correlation matrix and its many facets. Both introductory and advanced courses in statistics and quantitative methods focus carefully on the correlation coefficient, and some of those also discuss the correlation matrix as it pertains to advanced modeling methods. However, there is no comprehensive, dedicated, and unified treatment of the correlation matrix. The assumption might be that if students understand the correlation coefficient, they'll understand the correlation matrix. In fact, this assumption would be entirely incorrect. The goal of this book is to present the correlation matrix and many of its valuable details—and to do so assuming the least amount of mathematical background possible.

Teachers, students, and researchers in many disciplines will find this book valuable. The first disciplinary arena is obvious—that would be statistics, in both its theoretical and applied forms. But all disciplines that rely on quantitative analysis to support their research mission will find an introduction to the correlation matrix to be a valuable contribution to their scholarly repertoire. These include psychology, sociology, education economics, political science, business, communications, social work, anthropology, medical sciences, and biology, and there are many others. If a discipline includes an introductory course in statistics within its introductory curriculum, then this book will facilitate that effort.

The first author researched the correlation matrix for her master's thesis and made important breakthroughs (several of which are presented in Chapter 6); the seeds of this book were sown during that work in graduate school. The second author has been working on the theory and application of the correlation coefficient for his whole career; much of that treatment has contributed to statistical pedagogy. We continue a commitment to strong pedagogy within the current monograph. Why should we pay attention to the correlation coefficient, and how does it transition to the correlation matrix?

An argument can be made that the "invention" of the correlation coefficient by Karl Pearson was the most important development in the field of statistics. The correlation measures the relationship between two variables, which moves statistics from a univariate to a bivariate perspective. But there's more! We can organize correlations among many variables into a correlation matrix, but a correlation matrix is much more than a matrix filled with correlations. This matrix reflects the bivariate relationships among all pairs of variables. It also reflects the trivariate relationships among all triples of variables. It also reflects the quadravariate relationships among all quadruples of variables. It also reflects . . . well, you get the idea. In other words, just as a correlation transitions to a bivariate perspective, the correlation matrix transitions to a multivariate perspective. This multivariate space is where researchers naturally construct their models.

And so the starting point for all multivariate analysis methods is the correlation matrix, which can accommodate any number of variables. The multivariate relationships captured in a correlation matrix provide an exciting platform, from which researchers can depart on remarkable and exciting research journeys. To push the metaphor a bit further, the correlation matrix platform is the beginning of the journey, one that may well take us to other exciting statistical and methodological places, ones involving factor analysis, multilevel modeling, structural equation modeling, hazards modeling, and many other new and exciting statistical methods. Each of those statistical methods starts with a correlation matrix (or, perhaps, the unstandardized version of a correlation matrix, a covariance matrix).

This book is an introduction to the correlation matrix. We attempt to maintain a relatively accessible mathematical level throughout. Occasionally, when we must step up a mathematical level, we carefully warn the reader that this is happening, certain sections can be skimmed (or even ignored), and learning about correlation matrices can still proceed apace. But the focus, throughout, is on a conceptual understanding of the correlation matrix. Our audience has been, first, introductory and more advanced students of statistics (and, of course, their teachers). Our second audience has been researchers themselves, who likely work with and use correlation matrices, but who may not truly appreciate the "work of art" that started with Pearson's correlation coefficient, and then realized its multivariate potential in the correlation matrix. Enjoy the journey!

R files for chapters 1 and 3, plus an online appendix which demonstrates the use of the functions, are available on a website for the book at: **study .sagepub.com/researchmethods/qass/hadd-understanding-correlation.**

ACKNOWLEDGMENTS

This book has been in development for several years and benefited from the careful attention of several others. We express our appreciation, first, to Barbara Entwisle and Helen Salmon, who have expertly guided our writing toward a useful, readable, and technically sound contribution to the statistical literature. Patrick O'Keefe, who (like the first author) recently earned his PhD from the Quantitative Methods program in Vanderbilt's Peabody College of Education and Human Development, made several excellent suggestions and caught more than one mistake in early drafts. Finally, the second author acknowledges the long-standing conversation about correlation coefficients—which began in 1974, and continues to this day—with his friend and colleague Alan Nicewander. The reader can look up the Rodgers and Nicewander (1988) publication in the reference list to see an early product of this conversation. Parts of the current work contain more recent products.

Several reviewers recruited by Sage made outstanding suggestions that improved the book in many ways:

- Ralph Carlson, *The University of Texas Rio Grande Valley*
- Rich Dixon, *Texas State University*
- Garett C. Foster, *University of Missouri–St. Louis*
- Sally Jackson, *University of Illinois at Urbana–Champaign*
- William G. Jacoby, *Michigan State University*
- Shawn J. Latendresse, *Baylor University*
- Peter V. Marsden, *Harvard University*
- Thomas Rotolo, *Washington State University*
- Rick Tivis, *Idaho State University*

ABOUT THE AUTHORS

Alexandria Hadd is an assistant professor of psychology at Spelman College in Atlanta, where she teaches courses on statistics and research methods to undergraduate students. She earned her master's and PhD in quantitative psychology at Vanderbilt University and her BS in psychology and mathematics from Oglethorpe University. Her master's thesis—titled "Correlation Matrices in Cosine Space"—was specifically on the properties of correlation matrices. She also researched correlations in her dissertation, which was titled "A Comparison of Confidence Interval Techniques for Dependent Correlations." At Vanderbilt, she taught introductory statistics and was a teaching assistant for a number of graduate statistics/methods courses. In addition to correlation matrices, her research interests include applying modeling techniques to developmental, educational, and environmental psychology questions. Her hobbies include hiking, analog collaging, attending art and music shows, and raising worms (who are both pets and dedicated composting team members).

Joseph Lee Rodgers is Lois Autrey Betts Chair of Psychology and Human Development at Vanderbilt University in Nashville. He moved to Vanderbilt in 2012 from the University of Oklahoma, where he worked from 1981 until 2012, and where he holds the title George Lynn Cross Emeritus Professor of Psychology. He earned his PhD in quantitative psychology from the L. L. Thurstone Psychometric Laboratory at the University of North Carolina (UNC), Chapel Hill, in 1981 (and also minored in Biostatistics at UNC). He earned undergraduate degrees in mathematics and psychology from the University of Oklahoma in 1975. He has held short-term teaching/research positions at Ohio State, University of Hawaii, UNC, Duke, University of Southern Denmark, and Penn. He has published six books and more than 175 papers and chapters on statistics/quantitative methods, demography, behavior genetics, and developmental and social psychology. His best-known paper, "Thirteen Ways to Look at the Correlation Coefficient," was published in *The American Statistician* in 1988. He is married to Jacci Rodgers, an academic accountant (and currently an associate dean of Peabody College at Vanderbilt), and they have two adult daughters;

Rachel works for an international development company in Washington, D.C., and Naomi is a PhD student in geology at the University of Southern California in Los Angeles. His hobbies include playing tennis and golf, reading, and music.

Chapter 1

INTRODUCTION

The correlation matrix is a row-by-column arrangement of a set of correlation coefficients. The rows and columns refer to specific variables, which are measured features of the people, animals, or entities that behavioral science researchers study. For example, four variables assessed on people may be height, intelligence, birthweight, and shyness; three variables assessed on animals may be cortisol levels, reaction time, and counts of observed behaviors; and three variables assessed on an entity (e.g., a school) may be the percent low income, teacher turnover rate, and average student performance. A correlation matrix indicates the linear association between each pair of variables, such that the same variables in the same order label both the columns and the rows of the correlation matrix.

However, a correlation matrix is much more than an arrangement of individual correlation coefficients. Dozens of careful treatments of the correlation coefficient itself—the elements of a correlation matrix—exist in the statistical literature (e.g., Chen & Popovich, 2002; Rodgers & Nicewander, 1988) and in both sophisticated and introductory statistics textbooks. But understanding and appreciating the correlation *matrix* requires rather more careful study and mathematical sophistication than is required to understand the correlation *coefficient*. Few treatments—at either the introductory or more advanced level—extend the pedagogy of correlations from the separate correlation coefficients to the overall integrated correlation matrix. The current book, directed toward students, researchers, and methodologists who need to understand and/or teach correlation matrices, aims to provide this treatment.

We begin with a brief review of the correlation coefficient and of the related measure, the covariance. Correlation and covariance provide the foundation for many statistical techniques used across social, behavioral, and biological science disciplines. They also appear often in engineering, medical research, operations research, the physical sciences such as physics and chemistry, and other disciplines. Because correlations and covariances are the starting points for many statistical procedures, any discipline that defines its methods through statistical analysis is likely to rely extensively on these two measures of relationship. We treat both correlations and covariances throughout, though we will emphasize the correlation and, thus, will typically refer only to the correlation in general treatment. We will make clear when we are treating one or the other specifically. We distinguish the correlation and the covariance later in this chapter.

The Correlation Coefficient: A Conceptual Introduction

The correlation coefficient describes the linear association between two variables. It answers the question, "When one variable decreases or increases, how does the other variable tend to decrease or increase?" Correlation coefficients range from -1 to $+1$; magnitudes greater in absolute value (closer to $+1$ or -1) indicate a stronger association. Positive values indicate that as one variable increases (decreases), the other variable tends to increase (decrease)—that is, a positive or direct relationship. Negative values indicate that as one variable increases, the other variable tends to decrease (and vice versa)—that is, a negative or inverse relationship.

There are a number of different types of correlation coefficients, each with the purpose of quantifying the linear relationship between two variables. One way to categorize correlation coefficients into a taxonomy is defined based on the measurement level (Stevens, 1946) of the variables involved. That taxonomy assesses whether one or both variables are categorical (nominal), ordinal, or quantitative (interval or ratio). Often, "correlation" is shorthand for the Pearson product–moment correlation coefficient, the most common correlation measure that is used when both variables are quantitative, measured at interval or ratio levels within the Stevens measurement level system. Other correlation coefficients have been defined as well. For two ordinal variables, Spearman's rho, Kendall's tau, and polychoric correlation could be used to measure the relationship. For two binary (dichotomous) outcome variables, the phi coefficient and the tetrachoric correlation are the appropriate measures of association. For one binary and one quantitative variable, the point-biserial and biserial correlation coefficients are appropriate correlation measures.

Another classification system is the one used by Chen and Popovich (2002), which distinguishes between parametric and nonparametric measures. This distinction typically involves the question of whether a normal distribution is assumed to underlie one or both variables. For example, the formulas for polychoric and tetrachoric correlations assume that a normal distribution underlies the nonquantitative variables of interest. On the other hand, a nonparametric correlation "requires fewer assumptions and does not attempt to estimate population parameters" (Chen & Popovich, 2002, p. 79). Many nonparametric correlations are computed from variables that are naturally categorical (e.g., bright vs. dark colors, urban vs. rural residence) and do not have any underlying quantitative distribution, normal or otherwise. The phi coefficient and Spearman's rho are examples of nonparametric correlations.

A third way to classify correlations is in relation to the original correlation measure, the Pearson correlation. Several of the correlations defined above are actually special cases of the Pearson correlation applied to nonquantitative variables. For example, given two ordinal variables, we can rank order their values (within each variable); if we apply the Pearson correlation formula to those rank orders, we compute a Spearman rho correlation coefficient. Similarly, suppose we have two variables, one a typical quantitative variable and the other a binary variable coded with 0 indicating one category and 1 the other. If we use the Pearson correlation formula on that coding scheme, we are computing a point-biserial correlation coefficient. The phi coefficient is also a special case of the Pearson correlation, defined using the Pearson formula for two variables, each measured as binary variables. On the other hand, the Kendall tau ordinal correlation, the biserial correlation, and the tetrachoric/polychoric correlations are defined using different formulas that are not special cases of the Pearson correlation.

There exist many different ways to interpret a correlation coefficient. Rodgers and Nicewander (1988) showed that the correlation coefficient can be interpreted as one of several special kinds of means (e.g., the mean of the standardized cross products, or as a geometric mean), a special case of covariance, a special kind of variance, the slope of the standardized regression lines, a cosine, a function of the angle between two regression lines, and through several additional trigonometric interpretations. Standard introductory statistical textbooks show how to do null hypothesis significance testing (NHST) using the correlation coefficient, and the correlation measure can also be interpreted as an effect size. To conclude our review of the correlation coefficient—the building block for the correlation matrix—we present one of the standard formulas for the correlation coefficient. There exist many formulas that are algebraically equivalent, but conceptually distinct. We use one that allows us easily to demonstrate its relationship to a measure of covariance. If we define two quantitative variables, X and Y, with means $\bar{X}$ and $\bar{Y}$, respectively, for N observations (i.e., we define N pairs of scores), then the Pearson correlation coefficient can be computed using the following formula:

$$r_{XY} = \frac{\Sigma(X - \bar{X})(Y - \bar{Y})}{\sqrt{\Sigma(X - \bar{X})^2 \, \Sigma(Y - \bar{Y})^2}} \qquad (1.1)$$

The Covariance

Earlier, we referred to the covariance as a measure similar to the correlation. The cleanest way to conceptualize the relationship between the correlation and the covariance is to consider the correlation as a standardized version of the covariance. In other words, the correlation can be viewed as a measure of relationship between standardized variables, whereas the covariance is the measure of relationship between the equivalent unstandardized (or raw) variables. To appreciate this distinction requires understanding unstandardized and standardized variables.

When scores on a variable are collected using a particular scale of measurement (e.g., intelligence quotient [IQ], with a mean of 100 and standard deviation of 16; a shyness scale, with a mean of 50 and a standard deviation of 5; or adult female height, with a mean of 65 inches and a standard deviation of 3 inches), we typically refer to those measures as raw scores, measured on an unstandardized variable. If a respondent has a score of 68 inches on the height scale, and a score of 92 on the IQ scale, it is meaningless to compare those two raw scores; in no sense does the respondent have 24 more units of IQ than of height, because of the different scales of measurement.

This incompatibility is easily adjusted using standardization. Standardized scores (also called z scores) are defined for a given variable by subtracting the variable's mean and dividing by the variable's standard deviation; standardized scores indicate how far above or below the variable's mean that score is in terms of standard deviation (SD) units. Thus, the standardized score associated with a height of 68 inches is $z_{height} = (68 - 65)/3 = 1$; this computation tells us that a height of 68 is 1 SD unit above the mean. The standardized score associated with an IQ of 92 is $z_{IQ} = (92 - 100)/16 = -0.5$; this computation tells us that an IQ of 92 is 0.5 SD units below the mean. At this point, these two measurement scales have been standardized and are now at least loosely comparable.

This development allows us to distinguish the correlation from the covariance. The correlation can be defined in relation to z scores. A mathematically equivalent form of the Pearson correlation formula in Equation (1.1) is the following formula:

$$r_{XY} = \frac{\sum z_X z_Y}{N-1} \tag{1.2}$$

Furthermore, the correlation has defined bounds of +1.0 and −1.0. The covariance, which has no defined bounds in general, depends on the scales of measurement of the two variables. The formula for the covariance between two variables, X and Y, is the following. Note the similarity between Equations (1.2) and (1.3).

$$\text{Cov}(X,Y) = \frac{\sum (X - \bar{X})(Y - \bar{Y})}{N-1} \tag{1.3}$$

In some research settings, it is important to define statistical procedures that account for the different scales of measurement of the variables. We will briefly touch on such issues when we discuss factor analysis and structural equation modeling (SEM) in Chapter 4. In such settings, covariances—and covariance matrices—are the preferred measures of association. In other settings, the researcher would prefer to equate the scales of measurement—using standardization—so that differences in the scales' means and standard deviations can be ignored. In those settings, correlations—and correlation matrices—are the preferred measures of association. There is no correct answer to the question, "Which should be used, a correlation (matrix) or a covariance (matrix)?" The answer depends on the researchers' goals and how the variables were measured. This book—ostensibly about correlation matrices—is also about covariance matrices as well. By the end of the book, the reader will have some insight into when correlation matrices are preferable to covariance matrices, and vice versa. We emphasize, however, that our typical (and default) treatment in this book is of the correlation matrix.

The Correlation Coefficient and Linear Algebra: Brief Histories

It is not coincidence that the two developers of the correlation coefficient—Francis Galton and Karl Pearson, in the late 1800s—were collectively interested in a wide range of scientific disciplines, including psychology, genetics, geography, astronomy, sociology, and biometrics (Stanton, 2001). These fields required a measure that would appropriately capture the association between two quantitative variables. Galton first proposed the idea of the correlation coefficient, stemming from his earlier work on regression (Galton, 1885), while conducting research on the correspondence between parents' and their offspring's physical traits. Through this work, he realized there existed an "index of correlation" that captured the linear association between heights in kinship pairs. By 1890, he understood that the idea of correlation extended beyond questions of heredity and could be applied broadly to any two quantitative variables—and not simply to measures of the same construct, as he had originally thought (Stigler, 1989). However, it was his student, Pearson, who developed the mathematical formula and theory of the product–moment correlation that is still used most commonly today (Pearson, 1896).

Preceding the development of correlation by only a few decades was the development of linear, or matrix, algebra. Linear algebra grew out of the study of determinants for systems of linear equations in the early 1800s. Determinants are measures obtained from a matrix that reflect the linear relationships inside the matrix and, thus, are mathematically related to correlations. Interestingly, the mathematical concept of determinants (now wedded to the mathematics of matrices) developed well before matrix algebra; determinants were referenced at least as early as the 17th century by Leibnitz. In 1848, J. J. Sylvester first used the term *matrix* in a mathematical setting, the word *matrix* deriving from Latin for "womb," "mother," or "place where something develops." In 1855, Arthur Cayley first referred to a matrix with a single, uppercase letter, thereby cementing matrices as entities more complete than and distinguishable from their separate elements. The first linear algebra textbook, appropriately titled *Linear Algebra* by Hüseyin Tevfik Pasha, was written by happenstance almost contemporaneously with the development of the correlation coefficient (though linear algebra developed in what is today Bulgaria, whereas correlation and regression developed largely in England).

The development of both matrix algebra and the correlation coefficient set the stage for the rapid development of the correlation matrix and statistical methods applied to the correlation matrix. Unsurprisingly, Pearson was one of the first psychometricians to incorporate the newly developed and quickly expanding field of matrix algebra into his conceptualization of the correlation coefficient. In his groundbreaking 1901 article, in which he proposed what would later become principal components analysis (PCA), Pearson demonstrated both the computation of a determinant and what a correlation matrix between q variables would look like (see Figure 1.1).

However, he did not refer to the mathematical entity he created as a correlation matrix (the term *matrix* does not appear in the article), nor did he consider the correlation matrix beyond its convenient notation for producing the determinant. Three years later, Spearman (1904), a psychologist who made extensive contributions to statistics (including early work on factor analysis and adapting Pearson's correlation formula for ordinal variables), published what may be the first empirical correlation matrix (except that the diagonal had been modified; Figure 1.2); these correlations relate measures among British schoolchildren of "talent" within these different "branches."

Figure 1.1   The Determinant of a Generic Correlation Matrix, Appearing in Pearson (1901)

$$\Delta = \begin{vmatrix} 1 & r_{12} & r_{13} \ldots\ldots r_{1q} \\ r_{21} & 1 & r_{23} \ldots\ldots r_{2q} \\ \cdot & \cdot & \cdot \quad \cdot \quad \cdot \quad \cdot \\ r_{q1} & r_{q2} & r_{q3} \ldots\ldots 1 \end{vmatrix} \quad \cdots\cdots \quad \textbf{(xvi.),}$$

Figure 1.2 A Modified Correlation Matrix, Appearing in Spearman
(1904, p. 275)

	Classics.	French.	English.	Mathem.	Discrim.	Music.
Classics,	*0.87*	0.83	0.78	0.70	0.66	0.63
French,	0.83	*0.84*	0.67	0.67	0.65	0.57
English,	0.78	0.67	*0.89*	0.64	0.54	0.51
Mathem.,	0.70	0.67	0.64	*0.88*	0.45	0.51
Discrim.,	0.66	0.65	0.54	0.45		0.40
Music,	0.63	0.57	0.51	0.51	0.40	

Similar to Pearson, Spearman (1904) did not refer to his table as a "correlation matrix" but rather a "table of correlation," with instructions to the reader for how to read the table: "Each number shows the correlation between the faculty vertically above and that horizontally to the left" (p. 274).

Correlation matrices in scholarly literature had a small but consistently increasing number of mentions in the decades following the contributions of Spearman and Pearson. However, in recent decades, the explicit use of correlation matrices has increased exponentially. According to an informal search of the ProQuest online scholarly text database, there were 32 peer-reviewed records in the 1930s that mentioned the term *correlation matrix*. By the 1980s, the number of records grew to 1,827. By the most recent count for the 2010s that number is 31,505, with references in mathematics, neuroimaging, environmental science, applied psychology, and business journals. The wealth of attention to correlation matrices in applied research is likely due to a conflux of multiple factors, including modern computation, advances in data collection techniques, and advances in methods to analyze correlation matrices.

We provide here a brief summary of the history of correlation matrices to support what is well-known among statisticians, but which is less obvious for novice statistical students: Correlation matrices are more than just the convenient square arrangement of correlation coefficients. Inspection of correlation matrices facilitates a deeper understanding of the *multivariate* relationships among variables and allows for more complex theory development and testing than can possibly emerge from inspection of the separate disjoint correlations. One way to begin to appreciate the nuance of a correlation matrix is to recognize that not only does a correlation matrix include information about pairs of variables, but it also implicitly contains mathematical constraints that apply to relationships among triples of variables, to quadruples of variables, and so on. This book is written to develop intuition and understanding for correlation matrices, the tests that may be conducted on them, the modeling that can be applied to them, and the graphical methods that may be used to display them.

Examples of Correlation Matrices

In the following paragraphs, we develop a number of examples of correlation matrices (and, in several cases, the equivalent covariance matrix as well). These examples are based on real data collected in real research settings. They are chosen to be disciplinarily broad, including variables that would be used in education, psychology, sociology, political science, economics, communications, health care research, and other social/behavioral sciences settings. Once defined, we use these specific correlation matrices throughout the chapters of this book to illustrate principles and application of statistical methods relevant to correlation matrices.

As an example of how correlation matrices can motivate hypotheses or empirical analyses that are difficult to interpret using only bivariate correlations, consider Tables 1.1 and 1.2 adapted from Humphreys et al. (1985). Each table represents a correlation matrix capturing how a construct (intelligence of boys and girls, respectively) correlates within person over development. Although inspection of any given element of the correlation matrix would indicate that boys' (or girls') intelligence at one time is positively associated with intelligences at another time, the correlation matrix structure makes salient that intelligence measurements closer together are more strongly correlated than those measured further apart. Furthermore, the patterns of correlations are similar for boys and girls, and using methods presented in this book, we can formally test if the correlations are equivalent across gender in the population. In addition, at younger ages intelligence does not correlate as strongly with adjacent time points as it does at later time points. These kinds of observations would not be either obvious or easy to discuss if we relied on inspection of separate correlation coefficients. Within the context of a correlation matrix, their description and study are straightforward.

Although the correlations in Tables 1.1 and 1.2 were calculated on individuals (i.e., children across years of development), correlation matrices are also frequently used to show relationships for which the unit of analysis is a group. As examples, we have included two correlation matrices based on groups. Table 1.3 demonstrates a correlation matrix, adapted from Elgar (2010), calculated from 33 countries for which the variables of interest are country-level indicators of income inequality, average tendency to trust others, public health expenditures, life expectancy, and adult mortality.

Table 1.1 Correlations Between Boys' Intelligence Measured at Different Ages

Age (Years)	8	9	10	11	12	13	14	15	16	17
8	1.00	.60	.63	.67	.64	.59	.60	.62	.62	.53
9	.60	1.00	.74	.70	.68	.68	.68	.59	.60	.57
10	.63	.74	1.00	.79	.77	.70	.75	.71	.71	.60
11	.67	.70	.79	1.00	.87	.78	.75	.79	.81	.75
12	.64	.68	.77	.87	1.00	.84	.79	.77	.80	.76
13	.59	.68	.70	.78	.84	1.00	.85	.77	.77	.77
14	.60	.68	.75	.75	.79	.85	1.00	.84	.80	.75
15	.62	.59	.71	.79	.77	.77	.84	1.00	.88	.78
16	.62	.60	.71	.81	.80	.77	.80	.88	1.00	.85
17	.53	.57	.60	.75	.76	.77	.75	.78	.85	1.00

Note: The longitudinal sample of boys from the Boston area used a variety of measures of intelligence across the 10 yearly time points. Correlations in the original article were calculated pairwise, with sample sizes differing between 391 and 511; for examples throughout the book, we use a conservative N of 391. Adapted from Humphreys et al. (1985).

Table 1.2 Correlations Between Girls' Intelligence Measured at Different Ages

Age (Years)	8	9	10	11	12	13	14	15	16	17
8	1.00	.67	.64	.70	.69	.64	.64	.64	.63	.54
9	.67	1.00	.65	.68	.73	.73	.69	.61	.61	.61
10	.64	.65	1.00	.78	.78	.73	.73	.73	.69	.59
11	.70	.68	.78	1.00	.88	.80	.79	.80	.80	.75
12	.69	.73	.78	.88	1.00	.85	.84	.79	.79	.77
13	.64	.73	.73	.80	.85	1.00	.85	.75	.77	.79
14	.64	.69	.73	.79	.84	.85	1.00	.81	.77	.75
15	.64	.61	.73	.80	.79	.75	.81	1.00	.90	.79
16	.63	.61	.69	.80	.79	.77	.77	.90	1.00	.87
17	.54	.61	.59	.75	.77	.79	.75	.79	.87	1.00

Note: The longitudinal sample of girls from the Boston area used a variety of measures of intelligence across the 10 yearly time points. Correlations in the original article were calculated pairwise, with sample sizes differing between 495 and 693; for examples throughout the book, we use a conservative N of 495. Adapted from Humphreys et al. (1985).

10

Table 1.4 presents correlations for vital statistics—well-being, population, income, life expectancy, and rate of firearm deaths—for the 50 states in the United States.

Note that because the correlation coefficient is symmetric (e.g., the correlation between healthy life expectancy and adult mortality is the same as the correlation between adult mortality and healthy life expectancy), the correlation matrix is also symmetric across the diagonal (more on this topic in Chapter 2). Therefore, only the upper-triangular half or lower-triangular half of the matrix need to be shown. In the examples in this chapter, we highlight several different common styles of presenting correlation matrices in scholarly literature. For example, in Table 1.3, we used only the lower-triangular half of the table to show the entire correlation matrix. In Table 1.4, we showed both triangles of the table, and it can be easily verified that the correlations are symmetric by comparing equivalent correlations (e.g., compare the correlation between the first and second variable to the correlation between the second and first; both correlations equal .050).

Table 1.3   Correlations Between Income Inequality, Country-Averaged Tendency to Trust Others, and Measures of Public Health

Variable	Income Inequality	Trust	Public Health Expenditures	Healthy Life Expectancy	Adult Mortality
Income inequality	1.00				
Trust	−.51	1.00			
Public health expenditures	−.45	.12	1.00		
Healthy life expectancy	−.74	.48	.34	1.00	
Adult mortality	.55	−.47	−.13	−.92	1.00
Mean	0.363	3.9	5.6	68.0	0.129
SD	0.769	1.1	2.1	6.2	0.090

Note: Data came from the International Social Survey Program and included 48,641 respondents across 33 countries. Income inequality was assessed using data from the World Bank World Development Indicators database. Trust was measured as a country-level average of participants' rating of the statement "There are only a few people that I can trust completely," defined on a 5-point Likert-type scale (1 = *strongly agree*, 5 = *strongly disagree*). Public health expenditures, healthy life expectancy, and adult mortality measures were accessed using data from the World Health Organization Statistical Information System. Adapted from Elgar (2010).

Table 1.4 State-Level Vital Statistics for 2016–2017

Variable	1	2	3	4	5
1. Well-being	1.00	.050	.422	.708	−.416
2. Population	.050	1.00	.129	.246	−.254
3. Per capita income	.422	.129	1.00	.748	−.682
4. Life expectancy	.708	.246	.748	1.00	−.803
5. Firearm death rate	−.416	−.254	−.682	−.803	1.00
Mean	61.5	645	30.5	78.2	13.6
SD	1.20	720	4.24	1.76	5.32

Note: Well-being was measured using the Gallup-Sharecare Well-Being Index. Population was measured in units of 10,000 per estimates from the U.S. Census Bureau in 2016. Per capita income in units of $1,000 was reported for 2017 per the *Chronicle of Higher Education*. Life expectancy is reported for 2017 by National Geographic. Firearm death rate was reported for 2017 per the Centers for Disease Control and Prevention.

In each of Tables 1.3 and 1.4, we included rows showing the mean and standard deviation for each variable. We provided the additional information in these tables—and also in several future tables—so that readers can transform the correlation matrix into a covariance matrix. The formulas to transform a correlation matrix into a covariance matrix (and back again) rely on matrix algebra and are beyond the scope of treatment in the current book. We show, however, the relationship between a single correlation and its equivalent covariance:

$$\text{Cov}(X, Y) = r_{XY} * SD_X * SD_Y \tag{1.4}$$

This transformation requires the correlation and the standard deviations of the two variables. To provide a computational example, consider the correlation between Well-Being (WB) and Population (POP) in Table 1.4, $r = .050$. Using Equation 1.4, the covariance can be computed to be $\text{Cov}_{WB, POP} = .050*1.20*720 = 43.2$. The covariance between Well-Being and Per Capita Income (IN) is $\text{Cov}_{WB, IN} = .422*1.20*4.24 = 2.15$. Just as unstandardized variables cannot be compared with one another (see the height–IQ example above), covariances also cannot be compared. But because covariances transformed to correlations adjust out measurement scale differences and thus become comparable, correlations can be compared with one another in a meaningful way. Thus, we can report that the relationship between Well-Being and Population ($r = .050$) is substantially weaker than the relationship between Well-Being and Per Capita Income ($r = .422$).

The two covariances (43.2 and 2.15), on the other hand, reflect both relationship differences and differences between the measurement scales and, thus, are not naturally comparable.

Consider also the correlations presented in Table 1.5, adapted from Leonard (1997), which summarize how the racial composition of a sports team is associated with the racial composition of the city that team represents for professional basketball (National Basketball Association [NBA]), football (National Football League [NFL]), and baseball (Major League Baseball [MLB]). The table is organized such that three correlation matrices (one each for basketball, football, and baseball) are collated side by side and presented with only the upper-triangular half of the matrix. Organization of the correlation coefficients into matrices, and then concisely displaying these correlation matrices simultaneously, facilitates identification of patterns in the data. For instance, although there are near-perfect correlations between the percentages of Black residents in a franchise city in 1980 and 1990 for all three sports, indicating consistency in percent minority between 1980 and 1990 for the cities in the sample, the correlations are different across the sports for the number of Black teammates on teams between 1983 and 1989. We also see that professional baseball demonstrates consistently lower correlations between racial composition of a team and racial composition of the franchise city than do professional basketball and football. Is there evidence that the associations between racial composition of cities and teams in baseball act differently from those in basketball and football in this time period? Is there evidence that the associations between racial composition of cities and teams in basketball act similarly to football? These hypotheses are most efficiently demonstrated and tested in the context of correlation matrices, rather than through the tedious and inefficient inspection of individual correlations.

The statistical methods for analyzing correlation matrices are useful for exploring how—and sometimes why—variables are intercorrelated. For example, consider the correlation matrix presented in Table 1.6 on childbearing intentions and outcomes (where we present the whole symmetric correlation matrix). The correlation matrix is slightly modified from data collected from the 1979 National Longitudinal Survey of Youth (NLSY79), a nationally representative sample that was first assessed in 1979 when youth were 14 to 22 years old; the sample has been followed at least biennially thereafter. A researcher may be interested in the roles of childbearing intentions and previous childbearing outcomes in predicting future childbearing outcomes. Methods like path analysis, or more generally SEM, are designed to investigate these underlying processes.

13

Table 1.5 Correlation Matrices Indicating the Association Between the Number of Black Members of a Professional Sports Team (No. Black) for a City in a Given Year and the Percentage of Black Residents of That City (% Black) in a Given Year by Professional Sport

Variable	No. Black 1983			No. Black 1989			% Black 1980			% Black 1990		
	NBA	NFL	MLB	NBA	NFL	MLB	NBA	NFL	MLB	NBA	NFL	MLB
No. Black 1983	1.00	1.00	1.00									
No. Black 1989	.41	.13	.09	1.00	1.00	1.00						
% Black 1980	-.06	-.05	-.18	.37	.36	.11	1.00	1.00	1.00			
% Black 1990	-.06	-.10	-.23	.29	.30	.04	.99	.99	.96	1.00	1.00	1.00
Mean	6.26	13.2	6.69	8.04	17.8	4.81	26.8	28.6	30.6	27.5	30.7	31.9
SD	1.79	2.90	2.78	1.89	4.23	2.02	20.5	20.1	17.7	21.3	19.4	18.6

Note: Sample sizes differed by year and professional sport; for all examples in later chapters using these data, we use a sample size of $N = 26$ teams. NBA = National Basketball Association; NFL = National Football League; MLB = Major League Baseball.

Other statistical methods for analyzing correlation matrices, such as factor analysis and PCA, can be used to construct novel measures and assessments. For example, the correlation matrix (presented in its lower-triangular form) in Table 1.7 shows nine questionnaire items measured on 6,007 individuals from the NLSY79. Four of the items appear to assess positive self-esteem, three items appear to measure risk taking that may result in positive change, and the remaining two items appear to measure openness to others and new experiences. In Chapter 4, we will discuss how statistical methods can be used to construct scales to more completely explore if these items measure what we presume they measure.

Correlation matrices not only are useful for testing novel hypotheses but also can be vital to exploring patterns that might not be detected in the raw data, especially when the number of variables is exceedingly large. For example, consider the correlation matrix of index components for Standard & Poor's 500 Index, which captures the performance of 500 leading U.S. businesses and serves as a metric for how U.S. stocks are performing. Variables from these 500 companies may be difficult to summarize, let alone envision, in their raw form. Approaches such as factor analysis, PCA, and graphical methods (all of which we develop in future chapters) can help shed light on the complex associations among very large correlation matrices of this type.

Table 1.6   Correlations Between Childbearing Intentions and Childbearing Outcomes for 7,000 NLSY79 Respondents

Variable	1	2	3	4	5
1. Ideal number of children (1979)	1.000	.876	−.020	.484	.114
2. Expected number of children (1979)	.876	1.000	−.487	.389	.023
3. Number of children (1980)	−.020	−.487	1.000	.120	.441
4. Ideal number of children (1982)	.484	.389	.120	1.000	.207
5. Number of children (2004)	.114	.023	.441	.207	1.000
Mean	2.53	2.36	0.143	2.40	1.98
SD	1.53	1.46	0.45	1.36	1.46

Note: Ideal number of children was truncated at 5+ children. Expected number of children was truncated at 4+ children. Total number of children born to the respondent was reported in 1980 and 2004. Polychoric correlations were calculated on each variable, and missing data were pairwise deleted. NLSY79 = 1979 National Longitudinal Survey of Youth.

Table 1.7 Correlations Between Survey Items in the NLSY79
($N = 6,007$)

Variable	1	2	3	4	5	6	7	8	9
1. I am a person of worth.	1.0								
2. I have a number of good qualities.	.73	1.0							
3. I have a positive attitude with myself and others.	.53	.57	1.0						
4. I am satisfied with myself.	.45	.47	.63	1.0					
5. Willing to take risks in occupation	.06	.07	.06	.04	1.0				
6. Willing to take risks in other people	.05	.05	.03	.03	.37	1.0			
7. Willing to take risks in making life changes	.03	.05	.03	.01	.50	.39	1.0		
8. Extraverted, enthusiastic	.12	.11	.17	.13	.05	.05	.08	1.0	
9. Open to new experiences, complex	.10	.10	.13	.11	.08	.03	.10	.29	1.0
Mean	1.5	1.4	1.6	1.8	3.9	4.1	4.2	5.0	5.2
SD	0.7	0.6	0.7	0.7	3.2	2.9	2.9	1.8	1.6

Note: Items 1 to 4 were measured from 1 (*strongly agree*) to 4 (*strongly disagree*) in 2006 and reverse coded for ease of interpretation. Items 5 to 7 were measured from 0 (*unwilling to take any risks*) to 10 (*fully prepared to take risks*) in 2010. Items 8 and 9 were measured from 1 (*strongly disagree*) to 7 (*strongly agree*) in 2014. NLSY79 = 1979 National Longitudinal Survey of Youth.

Summary

This book is about correlation matrices. We focus on helping the reader develop an appreciation and intuition for using correlation matrices. We are writing for the student or researcher in the social and behavioral sciences, but the contents will appeal to students and researchers in any discipline that uses correlations, and of course to statisticians and applied mathematicians as well. We do not assume advanced mathematical knowledge; an introductory graduate (or even undergraduate) course in statistics will suffice to get started with the material. We avoid references to advanced mathematical discourse, except in the few cases we believe advanced mathematics are necessary to understand the material we are presenting (and in those treatments, we are careful to warn mathematically less sophisticated students that the material may be skipped or scanned without loss of continuity). Researchers interested in deeper treatment of matrices in general are referred to a linear or matrix algebra textbook. Also outside of the scope of this particular book is deep treatment of statistical methods to analyze correlation matrices. Whole courses and many introductory and advanced textbooks are devoted to these methods, such as factor analysis (e.g., Finch, 2019; Kim & Mueller, 1978a, 1978b); PCA (Dunteman, 1989); SEM (Long, 1983; Preacher et al., 2008); and meta-analysis (Wolf, 1986). Each of these methods involves fitting models to correlation (or covariance) matrices, and we briefly review those at a conceptual level; we also refer readers who wish for more advanced treatment to appropriate references, such as those mentioned above, which are all available in the Sage Quantitative Applications in the Social Sciences (QASS) series.

The organization of this book is as follows. In Chapter 2, we explore mathematical properties of correlation matrices. We minimize throughout, the use of equations or sophisticated mathematical operations (e.g., advanced matrix algebra). This chapter is pivotal for understanding the structure and function of correlation matrices. In Chapter 3, we provide details on common null hypothesis significance tests for correlation matrices, including how to conduct these tests. In Chapter 4, we overview methods that use correlation matrices as the raw data, including factor analysis, SEM, and meta-analysis. In Chapter 5, we demonstrate graphical methods for displaying correlation matrices of varying sizes and structures, usually with reference to the correlation matrices that have been presented as examples in the current introductory chapter. In Chapter 6, we describe work on the geometric underpinnings of correlation matrices, which is where most of the recent modern study of correlation matrices in the statistical literature has been focused. This chapter can be skimmed or skipped by introductory students. Finally, in Chapter 7, we provide a short conclusion and summary of the book.

Chapter 2

THE MATHEMATICS OF
CORRELATION MATRICES

Numbers are basic mathematical elements. Most people have a good intuition for what a number is. Seven is a number, as is -182.1. π (the ratio of a circle's circumference to its diameter, approximately 3.14159) is also a number. There are different types of numbers, such as integer, rational, irrational, and real numbers. Algebra and more advanced number theory textbooks can be consulted to develop understanding of numbers.

Numbers have characteristics. For example, a number can be whole or a decimal. A number can be positive or negative, real or imaginary. The number 2 is even; the number 7 is prime. Integer numbers are either even or odd. Even integer numbers bigger than 2 cannot be prime. Knowing the characteristics of a number helps us better understand the number.

Matrices are mathematical elements, like numbers. Matrices are defined as a rectangular array of numbers (called "scalars" in matrix algebra) arranged into rows and columns. Several examples of a certain type of matrix— correlation matrices—have already been presented in Chapter 1. The scalars that compose the matrix rows and columns can be seen in those examples; for example, in Table 1.4 there are 25 scalars, each representing a correlation between variables that combine to define a 5×5 correlation matrix.

Like numbers, matrices also have characteristics. One characteristic of a matrix is its dimensions—how many rows and columns the matrix has. By convention, the number of rows is listed before the number of columns; a 4×3 matrix is one that has four rows and three columns, for a total of $4 \times 3 = 12$ scalar elements. Some matrices with special patterns of dimensions have specific names. For example, a vector is a matrix that has either a single column or a single row; conceptually, the reader may imagine that the scalars are strung into a line horizontally (a "row vector") or vertically (a "column vector"). A square matrix is a matrix that has the same number of rows and columns; all correlation matrices are square matrices.

Each square matrix has a set of numbers associated with it called its eigenvalues that further characterize the matrix. Eigenvalues have different interpretations depending on the matrix (e.g., if it is a correlation matrix or some other type of matrix) and the field of study; they can be interpreted geometrically, whereby a matrix is related to an object in space, and they can be interpreted algebraically, whereby they relate to how the matrix changes when it is multiplied by itself many, many times. We will discuss eigenvalues only as they relate to correlation matrices, as they tend to be used in social science data analysis.

To provide an example of eigenvalues, consider the correlation matrix shown in Table 1.2 of girls' intelligence across development. The eigenvalues of this correlation matrix (there are 10, the same as the number of variables) are 7.63, 0.64, 0.41, 0.38, 0.25, 0.22, 0.18, 0.12, 0.10, and 0.07. The correlation matrix describing racial composition of cities and their corresponding NBA teams (Table 1.5) has four eigenvalues: 2.18, 1.32, 0.49, and 0.01. Eigenvalues cannot be calculated from a single correlation, or a subset of the correlations in a correlation matrix. The entire matrix is needed to determine what the eigenvalues of a correlation matrix are; they are determined using an eigenvalue formula.

In this chapter, we will describe why eigenvalues are an essential part of understanding correlation matrices. We will also refer to the practical significance of eigenvalues for methods to test null hypotheses about eigenvalues (Chapter 3) and to analyze correlation matrices (Chapter 4). We cover eigenvalues only as they are relevant to an applied researcher who uses correlation matrices. We skip the details of the calculation of eigenvalues, leaving those for linear algebra textbooks and software systems such as R, SAS, SPSS, and Stata. Math packages or online utilities will also readily compute the eigenvalues for a given matrix. Although formulas are basic arithmetic, the eigenvalue formula for a large correlation matrix (even a 10×10 correlation matrix can be considered fairly large in this context) requires a great deal of computational effort. Thus, computer software is virtually always used to compute eigenvalues. In this chapter, we will also present a summary of several other important mathematical features of a correlation matrix (some of these as they relate to eigenvalues), as well as notation that will be used throughout the rest of the book.

Requirements of Correlation Matrices

To understand a correlation matrix, it helps to start with the correlation coefficients themselves, which collectively define a correlation matrix. Correlations among a set of variables (e.g., $X_1, X_2, \ldots, X_p$) are typically summarized in a correlation matrix, which we will generically call $\boldsymbol{R}$. $\boldsymbol{R}$ will always be a square matrix of order p ("square" meaning that the matrix has the same number of rows as it has columns, and "order" here refers to the number of rows and columns, i.e., the number of variables); the rows and columns of $\boldsymbol{R}$ indicate the same p variables, those being correlated, and the entries in $\boldsymbol{R}$, r_{ij}, are correlation coefficients between pairs of variables X_i and X_j. For example, if the element of $\boldsymbol{R}$ in the fourth row and seventh column is $r_{47} = .24$, that would indicate that the correlation between variables X_4 and X_7 is .24. Note that, because r_{47} necessarily equals r_{74} (i.e., the correlation coefficient is a symmetric measure), the correlation matrix is symmetric.

The upper and lower triangles of a correlation matrix are defined in relation to the diagonal elements—that is, the elements in the r_{ii} positions in the matrix from the upper left to the lower right of the matrix. (Note that all diagonal elements, r_{ii}, in correlation matrices equal 1.0, because a variable correlates perfectly with itself.) The upper triangle consists of all elements in a correlation matrix that are above the diagonal; the lower triangle consists of all elements below the diagonal.

Elements r_{ij} in a correlation matrix—the correlations themselves—are constrained in the following four ways. Any researcher who has seen a correlation matrix, or studied the basic correlation coefficient, is likely familiar with the first three requirements:

1. $r_{ij} = r_{ji}$ (i.e., $\boldsymbol{R}$ is symmetric; the correlation between X_i and X_j is the same as the correlation between X_j and X_i)

2. $r_{ij} = 1$ if $i = j$ (i.e., the diagonal elements of $\boldsymbol{R}$ are 1; the correlation of a variable with itself is 1)

3. $-1 \leq r_{ij} \leq 1$ if $i \neq j$ (i.e., the off-diagonals of $\boldsymbol{R}$ are correlation coefficients bounded inside the interval between -1 and $+1$)

It is important to emphasize that these features of the correlation coefficient are simply mathematical properties of the formula by which the Pearson (and other) correlation coefficients are computed. Algebraic treatment that is relatively simple (but which we do not present here) exists to show that each of these properties is necessarily a feature of the correlation coefficient itself and, therefore, of all elements of a correlation matrix.

There is one final requirement for a correlation matrix that involves eigenvalues. The eigenvalues are related to the variances of the variables on which the correlation matrix is based; that is, the p eigenvalues are related to the variances of the p variables. True variances must be nonnegative, because they are computed from sums of squares, which themselves are each nonnegative. Thus, the final requirement for a correlation matrix is a check on its eigenvalues. Specifically, the fourth (and final) requirement is that all the eigenvalues of the correlation matrix are nonnegative:

4. $\lambda_1, \lambda_2, \ldots \lambda_p \geq 0$ where λ_i, $i = 1, 2, \ldots, p$ are the eigenvalues of $\boldsymbol{R}$.

We will discuss eigenvalues of a matrix in conceptual detail in the next section. Here, it is worth mentioning that this fourth requirement is often framed in terms of another characteristic of matrices, the matrix determinant, rather than eigenvalues. (Specifically, for readers with some matrix algebra background, this fourth property is equivalently ensured if the determinant of R and all principal minor submatrices of R are nonnegative.) We prefer to frame the fourth property in terms of eigenvalues, because eigenvalues may be calculated directly from R; the determinant equivalency requires the calculation of determinants from an increasingly large group of submatrices of R as p gets large.

Eigenvalues of a Correlation Matrix

Every square matrix has a set of eigenvalues and an associated set of eigenvectors. These are defined by mathematical definition, using specific formulas that can be found in any linear algebra text, or online (but which we do not present in any detail in this book). The eigenvalues and eigenvectors of a matrix are linked—each eigenvalue has a corresponding eigenvector, and vice versa. If a square matrix is of order p (i.e., p rows and columns), then the matrix has p eigenvalues and p eigenvectors. There may be repeating values among this set of eigenvalues, but the number of eigenvalues, with duplications, will still be p. Furthermore, the sum of the eigenvalues is equal to the sum of the diagonal elements of the matrix. Therefore, in the case of correlation matrices, in which the diagonal elements all equal 1 (and therefore the sum of the diagonal elements is p), the sum of the eigenvalues for the correlation matrix will also equal p. As examples of this property, in the fourth paragraph of this chapter where two sets of eigenvalues are presented, it is easy to verify that the first set of eigenvalues, from a 10×10 matrix, add to 10; the second set of eigenvalues, from a 4×4 matrix, add to 4.

Eigenvalues and eigenvectors are frequently invoked in fields that use statistical analysis. Readers don't need to have deep understanding and appreciation of these mathematical terms to use correlation matrices, but some conceptual understanding is useful to explain why some matrices that appear to be correlation matrices are not correlation matrices. Furthermore, and of more substantive interest, eigenvalues and eigenvectors have geometric interpretations that allow researchers to reduce complex information into simpler summaries. For example, facial recognition researchers use eigenvalues and eigenvectors to summarize similarities between many faces into a much smaller set of "eigenfaces"; audio recognition researchers can construct similar "eigenvoices" to break down complex speech into simpler dimensions. Intelligence researchers often summarize the information in a whole battery of instruments measuring human abilities by using factor analysis to smooth out the redundancies (i.e., overlapping variance) across the many different measures. Methods such as PCA and factor analysis rely on eigenvalues and eigenvectors to develop component and factor models of a set of variables.

In the previous section, we indicated that p eigenvalues are related to variances that underlie the correlation matrix. More specifically, eigenvalues, relative to p, are measures that are related to proportions of variance. In a correlation matrix with one or a few large eigenvalues, relative to p, substantial redundancy is indicated among the variables; that is, many of the variables share a great deal of variance and thus map into a central construct or dimension (technically, these dimensions are often called "principal components"). For example, a correlation matrix of order 4 may have eigenvalues 2.8, 0.9, 0.2, and 0.1 (note that these four eigenvalues sum to four, as required). The presence of the relatively large first eigenvalue of 2.8 indicates that the variables share substantial common variance—roughly $2.8/4 = 0.7$, or 70% of the variance among all the variables may be expressed with a single linear combination of the four variables. Next, $0.9/4 = 23\%$ of the variance is accounted for by a second dimension—this second dimension is constructed to be unrelated (uncorrelated) to the first dimension. Thus, underlying the four variables with these eigenvalues is one dominant dimension and a second uncorrelated, less dominant, dimension, with very little variance accounted for by the third and fourth dimensions (around 7%), which means that these four variables can be (almost completely) summarized by two dimensions (or components, or factors).

Pseudo-Correlation Matrices and Positive Definite Matrices

The constraint on correlation matrices that all eigenvalues must be non-negative occurs because eigenvalues are related to variances. As noted in the examples above, a given eigenvalue divided by the sum of all eigenvalues gives the proportion of variance associated with the particular direction or dimension defined by the associated eigenvector. The presence of a negative eigenvalue would therefore indicate a negative proportion of variance, which is conceptually and mathematically intractable for statistical settings. However, it is not unusual that a matrix may *look* like a correlation matrix because each element of the apparent correlation matrix meets the first three requirements of a correlation matrix, but the overall matrix fails to meet the fourth requirement of nonnegative eigenvalues. We call such matrices pseudo-correlation matrices.

For example, consider the (apparent) correlation matrix presented in Table 1.6. One would not be able to tell upon naive inspection that this correlation matrix—constructed using real data and modified slightly[1]—is in

---

[1] Only one correlation was altered from the correlation matrix computed from real data using polychoric correlations: The correlation between ideal and expected children in 1979 was increased from .756 to .876.

fact a pseudo-correlation matrix. The eigenvalues of this correlation matrix are 2.26, 1.59, 0.65, 0.50, and −0.0049. Because the correlation matrix has a negative eigenvalue, it is not a true correlation matrix.

Matrices that do satisfy all four requirements are called true correlation matrices. All true correlation matrices have nonnegative eigenvalues; in the language of matrix algebra, these are referred to as positive semidefinite (PSD) matrices. Correlation matrices that have strictly positive (i.e., no negative or zero) eigenvalues are positive definite (PD) matrices. Pseudo-correlation matrices are referred to in this language as indefinite matrices, indicating the presence of at least one negative and one positive eigenvalue.

If a correlation matrix has one or more eigenvalues that are exactly 0, this/these eigenvalues correspond to directions or dimensions (related to the corresponding eigenvectors) that explain zero proportion of the variance in the original variables. This circumstance may happen in practice if there is linear dependence among the variables in the correlation matrix. For example, a researcher may unintentionally create one variable that is a linear combination of one or more of the other variables, such as by including as variables in the correlation matrix both the total score and the individual items (which are summed to create the total score) in the same research setting. Correlation matrices that have one or more zero eigenvalues, even though a true correlation matrix, are problematic for most statistical software, and the researcher who tries to analyze such a correlation matrix may receive an error message from the computer program. In such cases, the researcher can check the data to ensure that they were entered correctly, or the researcher may be able to identify one or more variables that caused a linear dependence and that can be removed from the analysis.

Pseudo-correlation matrices are not just of theoretical interest; researchers often and regularly may have to diagnose and deal with such matrices. How do pseudo-correlation matrices exist? A pseudo-correlation matrix may look like a true correlation matrix, but there does not exist a set of complete quantitative variables to which the Pearson correlation formula can be applied in a pairwise fashion to produce the matrix. A pseudo-correlation matrix may arise from real data under one of several conditions. First, if there exist missing data among the variables, a correlation matrix created using pairwise complete cases (i.e., computed from correlations between pairs of variables, but because of missing data patterns using different subsets of the observations) may have one or more negative eigenvalues. The correlation matrix based on complete cases of numeric data using the Pearson product–moment correlation formula will necessarily be a true correlation matrix, but many researchers calculate correlations using as many observations as possible for each pair of variables, resulting in correlations within a correlation matrix based on different sample sizes and based on different subsets of the total data set, which can lead to pseudo-correlation matrices.

Second, a pseudo-correlation matrix may occur if the variables used to construct the correlation matrix are not numeric/quantitative but are rather binary or ordinal, in which case a researcher may choose to use polychoric or tetrachoric correlation formulas to form the correlation matrix. Polychoric and tetrachoric correlations are calculated by assuming that the binary/ordinal variables that are being correlated are attempting to measure traits that are inherently normally distributed; although these types of correlations are recommended in many research settings, as the assumption of underlying normality is often reasonable, an entire correlation matrix populated by polychoric or tetrachoric correlations may not be PSD and may be a pseudo-correlation matrix.

Third, if the correlation matrix is the result of averaging more than one correlation matrix (such as may be done in two-stage meta-analysis), then there is no guarantee that the resulting correlation matrix is PSD. In all of these cases, if the correlation matrix is in actuality a pseudo-correlation matrix, warnings or errors are likely to be generated by the statistical software system used to analyze the correlation matrix. The correlation matrix presented in Table 1.6, for example, is a pseudo-correlation matrix, and trying to analyze it will likely cause an error message in software.

Although problematic in substantive research settings, pseudo-correlation matrices can inform quantitative methods. Recent work has focused on using pseudo-correlations to provide insight into a true correlation matrix of interest (Waller, 2016). Other work has dealt with statistical issues surrounding pseudo-correlation matrices in real-data settings (Bentler & Yuan, 2011; Higham, 2002), particularly with large correlation matrices where the relative risk of a correlation matrix being non-PSD is greater.

Smoothing Techniques

In cases where a pseudo-correlation matrix did not arise from error, and the researcher does not wish to remove variables or alter how the correlation matrix was calculated to amend the non-PSD matrix, smoothing techniques are recommended before proceeding with statistical analyses. The goal of a smoothing technique is to produce a true correlation matrix that closely approximates the pseudo-correlation matrix. Generally, programs that implement smoothing techniques take as input the pseudo-correlation matrix and allow the user to indicate their tolerance for how much change is allowed to smooth the pseudo-correlation matrix into a true correlation matrix. The program will then output a smoothed, PD true correlation matrix that is "close" to the provided pseudo-correlation matrix.

There are a variety of smoothing techniques that can be broadly sorted into three categories: (1) shrinking techniques, (2) simultaneous-variable techniques, and (3) single-variable techniques. Shrinking techniques simply reduce the magnitude of all correlations toward zero; after a sufficient amount of shrinking, the correlation matrix typically will be PSD or strictly PD. Simultaneous-variable techniques seek the true correlation matrix that minimizes the "distance" to the pseudo-correlation matrix, using a variety of definitions for how "distance" is measured. Most smoothing techniques are simultaneous-variable techniques, and several of these have been proposed (e.g., Rousseeuw & Molenberghs, 1993); however, techniques performed with comparable tolerance levels will provide similar smoothed correlation matrices (Kracht & Waller, 2018). Table 2.1 demonstrates two different simultaneous-variable smoothing techniques for the correlation matrix in Table 1.6. Both techniques were implemented in the free software program R and are functions in popular R packages. Finally, single-variable techniques focus on only shrinking rows and columns of the correlation matrix associated with one or a few "problem" variables (e.g., Bentler & Yuan, 2011).

Table 2.1 Two Smoothed, PD Correlation Matrices Calculated From the Non-PSD Matrix in Table 1.6 Using the Software Package R

Variable	1	2	3	4	5
1. Ideal number of children (1979)	1.000	*.763*	−.053	**.473**	.121
2. Expected number of children (1979)	*.762*	1.000	−**.425**	**.374**	**.010**
3. Number of children (1980)	−**.053**	−**.425**	1.000	.118	**.429**
4. Ideal number of children (1982)	**.473**	**.374**	.118	1.000	.207
5. Number of children (2004)	.121	**.010**	**.429**	.207	1.000

Note: Correlations above the diagonal were smoothed with the *nearPD()* function in the Matrix package, and correlations below the diagonal were smoothed with the *cor.smooth()* function in the psych package. The eigenvalues of both true smoothed correlation matrices are the same to two decimals: 2.15, 1.57, .65, .49, and .14. The smoothed matrices are identical to the third decimal except for one correlation (italicized). The two smoothing algorithms were implemented such that the smallest eigenvalue between the two procedures would be comparable. Correlations that have changed in magnitude by more than .10 from the corresponding element in the non-PSD matrix are bolded. PD = positive definite; PSD = positive semidefinite.

Single-variable techniques leave larger portions of the non-PSD correlation matrix unchanged but often result in greater change to the affected rows and columns compared with the simultaneous-variable methods.

Restriction of Correlation Ranges in the Matrix

We reiterate that a correlation matrix is not just a matrix filled with correlations. Not every set of correlations, arbitrarily inscribed symmetrically in the $p \times p$ matrix, will produce a true correlation matrix. Once one correlation coefficient in the matrix is known, or fixed, then other correlations in the matrix are constrained, or bounded, if a true correlation matrix is to be produced. Stanley and Wang (1969) derived the formula for the simple 3×3 correlation matrix case showing how fixing two of the correlation coefficients constrains the third correlation coefficient. That is, if a researcher has three variables of interest—say X_1, X_2, and X_3—and the values of r_{12} and r_{13} are known, the range of possible values for r_{23} can be mathematically derived and will generally be much tighter than the range of $[-1, +1]$. Hubert (1972) extended this formula for any number of variables. In Chapter 6, we show how this restriction of correlation range can be directly observed using the geometric representation of the set of all pseudo-correlation matrices.

The Inverse of a Correlation Matrix

Another characteristic of a matrix is its inverse. The inverse of a matrix is conceptually similar to the reciprocal of a scalar, the types of numbers that we deal with on a regular basis. For example, the scalars 8 and 36.9 have as reciprocals $1/8 = .125$, and $1/36.9 = \overline{.02710}$, respectively. Reciprocals or scalar inverses are the numbers that, when multiplied by the original number, produce the identity, 1.0. The identity scalar, 1.0, is the number that, when multiplied by any other number, returns the original number.

Although many scalar operations have equivalent operations on matrices (e.g., matrices of "matching" or conformable sizes can be added, subtracted, or multiplied), there is no matrix operation for division. For scalars a and b, you can simply calculate $\frac{a}{b}$, unless b is zero, in which case the ratio is mathematically impossible to calculate; for matrices A and B, it is impossible to calculate $\frac{A}{B}$, much like it is impossible to divide a scalar by 0. However, as for scalars, matrix inverses can be multiplied by other matrices as a substitute for division. For scalars, multiplying $a * \frac{1}{b}$ or $a * b^{-1}$, where $\frac{1}{b}$

or b^{-1} is the reciprocal of b, is the same as calculating $\frac{a}{b}$. For matrices, it may be possible to calculate AB^{-1}, where B^{-1} is the inverse of matrix B, even though $\frac{A}{B}$ can never be calculated.

There is a matrix identity that, when multiplied by another matrix, returns that original matrix, and we can define the matrix inverse as the matrix that, when multiplied by the original matrix, will equal the identity. The details of actually computing matrix inverses are not important to studying correlation matrices, but it is important to know that in the computations used to do statistical analysis, the inverse of the correlation matrix (R^{-1}) is often used rather than R itself. R^{-1} has direct interpretations in advanced statistical methods such as multiple regression, factor analysis, and discriminant analysis (Raveh, 1985) and, sometimes, also serves as a weight matrix in analysis. However, just as the scalar number zero has no reciprocal, certain matrices also do not have inverses, including, for example, pseudo-correlation matrices and PSD correlation matrices. Among correlation matrices, only true, strictly PD correlation matrices have inverses—which explains why many statistical programs will return an error message if the researcher tries to analyze a correlation matrix with one or more zero or negative eigenvalues. A cryptic message that "the correlation/covariance matrix cannot be inverted," or equivalently, "the correlation/covariance matrix is not full rank," is referencing the absence of a valid inverse for the correlation matrix.

The Determinant of a Correlation Matrix

The final characteristic of a matrix we find relevant to (briefly) discuss in this book is the determinant of a matrix. All square matrices have a determinant (denoted as $|R|$ for a given correlation matrix R), which is a single number equal to the product of all of the eigenvalues of the matrix. Computer programs can readily calculate the determinant of a matrix, along with the eigenvalues and eigenvectors. Although inspecting the p eigenvalues of R is often useful, the determinant can provide some quick diagnosis for issues about the correlation matrix. For example, if $|R| < 0$, then one or more of the eigenvalues of R is negative, and R is therefore a pseudo-correlation matrix. If $|R| = 0$, then one or more of the eigenvalues of R is equal to 0, and there is linear dependence among the variables of the correlation matrix. Finally, $|R| > 0$ for true correlation matrices that are PD and have eigenvalues that are strictly positive (although in some fairly unusual cases a pseudo-correlation matrix may have $|R| > 0$, such as if an even number of eigenvalues are negative). We discuss determinants primarily because they appear in some test statistics for null hypotheses on correlation matrices, which we discuss in Chapter 3.

Examples

Racial Composition of NBA and Sponsor Cities

The correlation matrix in Table 1.5 has four variables. For this correlation matrix, X_1 = number of Black teammates in 1983, X_2 = number of Black teammates in 1989, X_3 = percent Black of city residents in 1980, and X_4 = percent Black of city residents in 1990. This correlation matrix is obviously of order 4 ($p = 4$). Each entry in the correlation matrix is between $[-1, 1]$, and each element is a correlation coefficient. For example, $r_{12} = .41$ indicates that the correlation between the number of Black teammates on NBA teams between 1983 and 1989 is .41, or positively related at a moderate level. The value $r_{24} = .29$ indicates that there is also a positive (but weaker) relationship between the percentage of Black city residents in 1990 and the number of Black teammates on that city's NBA team.

The eigenvalues of this correlation matrix are 2.18, 1.33, 0.49, and 0.01 (which sum to 4 within rounding error). All the eigenvalues are positive, and so this matrix is strictly PD, and is therefore a true correlation matrix. The first eigenvalue, 2.18, is linked to an eigenvector that corresponds to a dimension accounting for 2.18/4 = .55, or about 55% of the total variance in the correlation matrix. The second eigenvalue corresponds to an eigenvector associated with an uncorrelated dimension that accounts for an additional 1.33/4 = .32 proportion of variance (and which is constrained to be uncorrelated with the first dimension). Therefore, two uncorrelated underlying dimensions corresponding to the first two eigenvectors have eigenvalues large enough to indicate that these two dimensions account for about 87% of the total variance among the four variables.

Girls' Intelligence Across Development

The correlation matrix in Table 1.2 has 10 variables—girls' intelligence measured each year from ages 8 to 17. The correlation matrix is obviously of order 10 ($p = 10$). The eigenvalues of this correlation matrix are 7.63, 0.64, 0.41, 0.38, 0.25, 0.22, 0.18, 0.12, 0.10, and 0.07 (which sum to 10 within rounding error). Because all the eigenvalues are positive, this correlation matrix is strictly PD and is, therefore, a true correlation matrix. The first eigenvalue is relatively large compared with the other eigenvalues (7.63/10 = 0.763), indicating that the first dimension is associated with a large portion of the variance (around 76%) among intelligence scores measured between ages 8 and 17.

28

Summary

Certain characteristics of correlation matrices are important for statistical applications. Correlation matrices contain correlations, but not all matrices that contain correlations are true correlation matrices. Those that appear to be correlation matrices by virtue of containing correlations, but are not true correlation matrices, are called pseudo-correlation matrices. How can we tell them apart? Pseudo-correlation matrices can be diagnosed by computing the eigenvalues that correspond to a particular correlation matrix (many software routines, or online computational applications, can be used to compute eigenvalues). True correlation matrices have eigenvalues that are only positive and/or zero. Pseudo-correlation matrices have at least one eigenvalue that is negative. Matrices with only positive eigenvalues are called PD matrices. Matrices whose eigenvalues are positive and/or zero are called PSD matrices. Matrices with at least one positive and one negative eigenvalue are called indefinite matrices; all pseudo-correlation matrices are indefinite. Smoothing techniques are algorithms that replace a pseudo-correlation matrix with the closest true correlation matrix, with "closest" defined differently across different techniques. In addition to eigenvalues, other important characteristics of the correlation matrix include its determinant and its inverse, both of which appear in statistical tests on correlation matrices described in the next chapter.

In the next chapter, we discuss a number of statistical procedures that have been developed to analyze correlation matrices. The material in the current chapter informs those analytic methods, because most of those approaches cannot be applied to pseudo-correlation matrices. The reader will see eigenvalues discussed in the next chapter (and also later in the book) and should by now be aware of their substantial value as diagnostic indices that reveal important features of both true and pseudo-correlation matrices.

Chapter 3

STATISTICAL HYPOTHESIS TESTING ON CORRELATION MATRICES

In this chapter, we review the literature devoted to statistical tests on correlation matrices and provide detailed recommendations and examples of a variety of hypothesis tests applied to correlation matrices. Hypothesis tests and confidence intervals for individual correlation coefficients are well-known; for example, a nil-null hypothesis test (i.e., where the null-hypothesized value is 0) on the associated regression coefficient for predicting Y from X (or vice versa) is equivalent to testing the null hypothesis that $\rho = 0$, or an asymmetric confidence interval can be constructed around r using the Fisher z transformation. Testing correlation coefficients calculated from two independent samples is also straightforward and is covered in other sources. However, hypothesis tests involving correlation coefficients within the same correlation matrix require different techniques, because the correlations are *dependent*—they are calculated on the same sample of individuals. Furthermore, testing hypotheses about overall correlation matrices, or comparing two or more correlation matrices statistically, requires hypothesis testing approaches that are multivariate rather than univariate.

This chapter is organized as follows: We begin by discussing hypotheses about correlations within a given correlation matrix, then discuss hypotheses about the entire correlation matrix, followed by several tests on two or more correlation matrices, and end with a test on the eigenvalues of a correlation matrix. For each hypothesis, we divide coverage of the associated test into the following: a description of when a particular hypothesis would be tested, the formal statement of the null hypothesis of the test, formulation of the test statistic for the recommended test, and a data example of the test being conducted. In this chapter, we use "hypothesis test" and "significance test" interchangeably, within the standard NHST framework. We use r_{ij} and ρ_{ij} for a sample correlation coefficient and its corresponding population value, R and P for a sample correlation matrix and its corresponding population correlation matrix, and λ_i and λ_i^* for a sample eigenvalue and its corresponding population value. In addition, we will remind the reader of the appropriate notation for correlation matrix characteristics, as described in Chapter 2, whenever a test makes reference to these quantities.

Software Package in R: In addition to providing examples for calculating the statistical tests by hand in the following sections, the authors have developed an R package with functions that will conduct these statistical tests. This suite of functions in R, which we call CorMatTests, is available for download for free from the online appendix to use in the R software environment. Also in the online appendix, we have provided documentation that explains how to use each function and examples of the functions' use.

Hypotheses About Correlations in a Single Correlation Matrix

Testing Equality of Two Correlations in a Correlation Matrix (No Variable in Common)

Consider the situation in which a researcher is interested in comparing the strength of the linear relationship between two variables, say, X_j and X_k, to the strength of the relationship between two other variables, say, X_h and X_m. For example, a researcher may wish to test if the relationship between high school GPA (grade point average) and time spent on schoolwork (we'll call this r_{jk}, likely a positive correlation) is stronger (i.e., of higher magnitude) than the relationship between college GPA and class attendance (r_{hm}, also likely a positive correlation), measured on the same group of individuals as they pass through the educational system. The point estimates for r_{jk} and r_{hm} may be used to rank the strength of the linear relationships in the samples; if r_{jk} and r_{hm} are calculated from two separate independent samples, standard confidence interval techniques can provide evidence to evaluate whether the sample correlations support equality of the equivalent correlations in the population. However, if only one sample is used to calculate both r_{jk} and r_{hm}, as in this example with the GPAs and academic habits assessed longitudinally over time, then these two correlations are no longer independent because they are measured on the same individuals. These two correlations would be elements of the same correlation matrix.

Methodologists have proposed several different statistical approaches to test correlations under these circumstances. We follow the suggestions of Steiger (1980) and Meng et al. (1992), who recommended a test based on a statistic dating back at least to the 1940s and later adapted to perform well in small samples by inclusion of Fisher's z transformation (Dunn & Clark, 1969). Fisher's z-transformation formula is $z = \frac{1}{2} \ln\left(\frac{1+r}{1+r}\right)$, and it is used to transform the inherently skewed distribution of a correlation coefficient to an approximately normal distribution of zs.

Null Hypothesis. $H_0 : \rho_{jk} = \rho_{hm}$ (two correlations in the same correlation matrix are equal in the population)

Test Statistic

$$Z = \sqrt{\frac{(N-3)}{2 - 2s_{jk,hm}}} \left(z_{jk} - z_{hm} \right),$$

where N is the sample size used to calculate the correlation matrix, and z_{jk} and z_{hm} are the Fisher-transformed r_{jk} and r_{hm}, respectively. $s_{jk,hm}$ is calculated as

$$s_{jk,hm} = \frac{\psi}{\left(1 - \bar{r}_{jk,hm}^2\right)^2}$$

such that $\bar{r}_{jk,hm}$ is the average of r_{jk} and r_{hm} and ψ is calculated as

$$\psi = \frac{1}{2} \bar{r}_{jk,hm}^2 \left(r_{kh}^2 + r_{jh}^2 + r_{jm}^2 + r_{km}^2 \right) - \bar{r}_{jk,hm} \left(r_{jh} r_{kh} + r_{kh} r_{km} + r_{jh} r_{jm} + r_{jm} r_{km} \right)$$
$$+ r_{jh} r_{km} + r_{kh} r_{jm}$$

Z is distributed approximately as a standard normal under the null hypothesis, so a one- or two-tailed p value can be calculated from a table of areas under the normal curve (or, equivalently, from a software system providing the appropriate value). Note that other correlations in the correlation matrix (r_{kh}, the correlation between variables X_k and X_h; r_{km}, the correlation between variables X_k and X_m, etc.) in addition to r_{jk} and r_{hm} enter into the calculation of the Z test statistic.

Example. Using the correlation matrix of girls' intelligence over development presented in Table 1.2, we will use a two-tailed test to test the null hypothesis that the correlation for girls' intelligence measured at ages 10 and 11 is equal to the correlation for girls' intelligence measured at ages 16 and 17. This hypothesis may be theoretically motivated by wishing to compare stability of intelligence across equally spaced time points. For this correlation matrix, we assume $N = 495$, which is the lowest reported sample size in the original article.

For this example, $r_{jk} = r_{43} = .78$ and $r_{hm} = r_{10,9} = .87$, so $z_{jk} = 1.045$, $z_{hm} = 1.333$, and $\bar{r}_{jk,hm} = .825$; $r_{jh} = r_{4,10} = .75$, $r_{jm} = r_{49} = .80$, $r_{kh} = r_{3,10} = .59$, and $r_{km} = r_{39} = .69$.

$$\psi = \frac{1}{2}(.825)^2(.59^2 + .75^2 + .80^2 + .69^2) - .825[(.75)(.59) + (.59)(.69)$$
$$+ (.75)(.80) + (.80)(.69)] + (.75)(.69) + (.59)(.80) = 0.0279$$

$$s_{jk,hm} = \frac{0.0279}{(1 - .825^2)^2} = 0.273$$

$$Z = \sqrt{\frac{(495 - 3)}{2 - 2(0.273)}}(1.045 - 1.333) = -5.294$$

Using $\alpha = .025$ in the lower tail, -5.294 is well beyond the -1.96 critical value; therefore, we reject the null hypothesis and conclude that the alternative hypothesis is supported, that the population correlation between intelligence measured at ages 10 and 11 is different from the population correlation between intelligence measured at ages 16 and 17.

Testing Equality of Two Correlations in a Correlation Matrix (Variable in Common)

This hypothesis test is nearly identical to the one above, except that the two correlations being tested share a variable in common. For example, a researcher may wish to test if the relationship between high school GPA and time spent on schoolwork (r_{jk}) is stronger than the relationship between high school GPA and class attendance (r_{jh}). Note that the subscripts for the correlations share a letter in common, indicating the common variable that is being correlated with two other variables. This hypothesis is a special case of the hypothesis discussed above, but the common variable simplifies the calculation of test statistic Z.

Null Hypothesis

$H_0 : \rho_{jk} = \rho_{jh}$ (two correlations in the same correlation matrix with a common variable are equal in the population)

Test Statistic

$$Z = \sqrt{\frac{(N - 3)}{2 - 2s_{jk,jh}}}(z_{jk} - z_{jh})$$

where N is the sample size used to calculate the correlation matrix, z_{jk} and z_{jh} are the Fisher-transformed r_{jk} and r_{jh}, respectively, and $s_{jk,jh}$ is calculated as

$$s_{jk,jh} = \frac{\psi}{\left(1 - \bar{r}_{jk,jh}^2\right)^2}$$

such that $\bar{r}_{jk,jh}$ is the average of r_{jk} and r_{jh} and ψ is calculated as

$$\psi = \frac{1}{4}\bar{r}_{jk,jh}^2\left(2\bar{r}_{jk,jh}^2 + r_{kh}^2 - 4r_{kh} - 1\right) + \frac{1}{2}r_{kh}$$

Z is distributed approximately as a standard normal under the null hypothesis, so a one- or two-tailed p value can be calculated from a z table.

Example. For this example, we will use the racial composition of sports teams' data presented in Table 1.5. The original Leonard (1997) article does not indicate exact sample sizes for the number of teams in either 1983 or 1989 seasons; for this example, we assume each sport had $N = 26$ teams, which is likely to be a (very slight) underestimate. We will test whether the correlation between the number of Black teammates on an NBA team in 1989 and the percentage of Black residents in the host city in 1980 ($r_{jk} = r_{23}$ = .37) is equal to the correlation between the number of Black teammates on an NBA team in 1989 and the percentage of Black residents in the host city in 1990 ($r_{jh} = r_{24}$ = .29). For this example, $\bar{r}_{jk,jh}^2 = \left(\dfrac{.37+.29}{2}\right)^2 = .1089$, $r_{kh} = r_{34}$ = .99, $z_{jk} = 0.388$, and $z_{jh} = 0.299$.

$$\psi = \frac{1}{4}(.1089)\left[2(.1089)+(.99)^2-4(.99)-1\right]+\frac{1}{2}(.99)=0.393$$

$$s_{jk,jh} = \frac{0.393}{\left(1-.1089\right)^2} = 0.494$$

$$Z = \sqrt{\frac{(26-3)}{2-2(0.494)}}\left(0.388-0.299\right)=0.429$$

Assuming $\alpha = .05$, and a two-tailed test, $Z = 0.429$ is less than the critical value of 1.96; therefore, we would fail to reject the null hypothesis that the correlations between Black teammates in 1989 and the percentage of Black residents in host cities in 1980 and 1990, respectively, are the same and could not conclude that the correlations are significantly different from each other.

Testing Equality to a Specified Population Correlation Matrix

Standard treatments of the correlation coefficient cover methods of testing if a single correlation coefficient is equal to a particular value (say, equal to zero) in the population, either by testing a null-hypothesized value or by constructing a confidence interval. However, it is often advantageous to test multiple—or all—correlations in a correlation matrix against hypothesized population values at the same time. For example, a researcher may wish to test if all correlations between the p variables taken in pairs—the number of correlations is equal to $\dfrac{p(p-1)}{2}$—in the matrix are equal to zero in the population simultaneously. Another common implementation of this test is in the context of longitudinal research: A researcher may wish to test if the correlations in the matrix are all equal to some nonzero values to determine if the correlations between measures taken at different time lags are consistent. In all cases, testing the whole correlation matrix at once is efficient. As p increases, the number of pairs of variables—the number of correlations—increases much faster, which not only makes conducting tests for each separate correlation cumbersome but also increases the overall probability of making a Type I error.

To conduct this test, we suggest the test statistic published in Jennrich (1970). Note that this test statistic requires the inverse of a matrix and matrix multiplication in its formula. We purposefully omit how to compute these quantities by hand, as modern software readily computes them, but we supply the output necessary for users of the test to check their work (and we describe some of these matrix algebra concepts in Chapter 2).

Null Hypothesis

$H_0 : \boldsymbol{R} = \boldsymbol{P}$ (the sample correlation matrix came from the $p \times p$ population matrix $\boldsymbol{P}$)

Test Statistic

$$\chi^2 = \frac{1}{2}\operatorname{tr}\left(\boldsymbol{Z}^2\right) - \operatorname{diag}'\left(\boldsymbol{Z}\right)\boldsymbol{T}^{-1}\operatorname{diag}\left(\boldsymbol{Z}\right)$$

In this test statistic, $\boldsymbol{Z} = \sqrt{N}\left(\boldsymbol{P}\right)^{-1}\left(\boldsymbol{R} - \boldsymbol{P}\right)$; $\boldsymbol{Z}^2$ indicates standard matrix multiplication of Z with itself; $\boldsymbol{T} = \boldsymbol{P} \odot (\boldsymbol{P})^{-1} + \boldsymbol{I}$; diag() are the diagonal elements of a matrix (organized into a column vector); tr() is the trace of a matrix (calculated by summing all of the diagonal elements of the matrix); N is the sample size of the correlation matrix $\boldsymbol{R}$; $(\boldsymbol{P})^{-1}$ is the inverse matrix of $\boldsymbol{P}$; $\odot$ denotes multiplication of corresponding elements (vs. standard matrix multiplication, which does not involve simple multiplication of corresponding elements); and $\boldsymbol{I}$ is the identity matrix of order p, which has 1s along the diagonal and 0s everywhere else. The test statistic is asymptotically distributed as χ^2 with $df = \dfrac{p(p-1)}{2}$.

Example 1. Using the racial composition of sports teams' data in Table 1.5, and again assuming $N = 26$, we will test the hypothesis that the MLB correlation matrix came from a specific population correlation matrix, given below as P. The population matrix is populated with correlations of $+.50$ and arbitrarily chosen (ideally, P is determined based on theory).

$$P = \begin{bmatrix} 1.00 & .50 & .50 & .50 \\ & 1.00 & .50 & .50 \\ & & 1.00 & .50 \\ & & & 1.00 \end{bmatrix} \quad R = \begin{bmatrix} 1.00 & .09 & -.18 & -.23 \\ & 1.00 & .11 & .04 \\ & & 1.00 & .96 \\ & & & 1.00 \end{bmatrix}$$

$$P^{-1} = \begin{bmatrix} 1.6 & -0.4 & -0.4 & -0.4 \\ & 1.6 & -0.4 & -0.4 \\ & & 1.6 & -0.4 \\ & & & 1.6 \end{bmatrix} \quad R - P = \begin{bmatrix} .00 & -.41 & -.68 & -.73 \\ & .00 & -.39 & -.46 \\ & & .00 & .46 \\ & & & .00 \end{bmatrix}$$

$$Z = \begin{bmatrix} 3.71 & -1.61 & -5.69 & -5.96 \\ -0.47 & 2.56 & -2.73 & -3.20 \\ -3.22 & -1.41 & 1.24 & 6.18 \\ -3.73 & -2.12 & 5.94 & 1.49 \end{bmatrix}$$

$$T = \begin{bmatrix} 2.6 & -0.2 & -0.2 & -0.2 \\ & 2.6 & -0.2 & -0.2 \\ & & 2.6 & -0.2 \\ & & & 2.6 \end{bmatrix}$$

$$\chi^2 = \frac{1}{2}\text{tr}\left(Z^2\right) - \text{diag}'\left(Z\right)T^{-1}\text{diag}\left(Z\right) = 89.19$$

The $\alpha = .05$ critical value for a χ^2 distribution with $df = \dfrac{4(4-1)}{2} = 6$ is 12.59. Because $\chi^2 > 12.59$, we reject the null hypothesis and conclude that R did not come from population correlation matrix P.

Example 2. Using the country-level data on public health and trust in government (Table 1.3) as an example, with $N = 33$ countries, suppose we wished to test the hypothesis that all pairwise correlations are equal to zero in the population; that is, public health expenditures, mortality, income inequality, and average tendency to trust others are not related in these countries. This hypothesis corresponds to a test of whether P (in the population) is a correlation matrix with 1s along the diagonal and 0s everywhere else (readers with cursory matrix algebra knowledge will recognize this as an *identity matrix*, I, which functions very similar in matrix algebra as the number 1 does in scalar algebra). When the null hypothesis for the test is independence, the formula for the test statistic χ^2 simplifies:

$$\chi^2 = \frac{1}{2}\operatorname{tr}\left(Z^2\right)$$

In this test statistic, $Z = \sqrt{N}\left(R - P\right)$, and tr() is the trace of a matrix. For this example,

$$Z = \begin{bmatrix} 0.00 & -2.93 & -2.59 & -4.25 & 3.16 \\ -2.93 & 0.00 & 0.69 & 2.76 & -2.70 \\ -2.59 & 0.69 & 0.00 & 1.95 & -0.75 \\ -4.25 & 2.76 & 1.95 & 0.00 & -5.28 \\ 3.16 & -2.70 & -0.75 & -5.28 & 0.00 \end{bmatrix}$$

$$Z^2 = \begin{bmatrix} 43.32 & -22.03 & -12.68 & -29.83 & 32.31 \\ -22.03 & 23.95 & 14.98 & 28.07 & -24.34 \\ -12.68 & 14.98 & 11.53 & 16.84 & -20.35 \\ -29.83 & 28.07 & 16.84 & 57.42 & -22.33 \\ 32.31 & -24.34 & -20.35 & -22.33 & 45.76 \end{bmatrix}$$

$$\chi^2 = \frac{1}{2}\operatorname{tr}\left(Z^2\right) = 90.99$$

The $\alpha = .05$ critical value for a χ^2 distribution with $df = \dfrac{5(5-1)}{2} = 10$ is 18.31. Because $\chi^2 > 18.31$, we reject the null hypothesis and conclude that at least some of the pairwise correlations in the matrix are nonzero in the population; that is, there is at least one linear relationship between the country-level variables in Table 1.3 at the population level. Note that rejecting this null hypothesis is a rather weak substantive result, simply suggesting that there is at least one nonzero correlation somewhere within the population correlation matrix.

Hypotheses About Two or More Correlation Matrices

Many researchers are interested in how relationships among variables differ for naturally occurring or experimentally designated independent groups. For example, we might be interested in how relationships differ among a whole set of variables for boys versus girls, for an experimental versus control group, for people from two different countries or regions, or for people of similar demographics but from different periods of time. Testing the equality of two or more correlation matrices can be a first step before more advanced analyses of how and why the relationships among variables differ between groups.

Several statistics have been developed to test the equality of two or more correlations. One of the first developed was by Kullback (1967) and assumed that the variables are approximately normally distributed. Kullback's test is very general, however, Jennrich (1970) shows that Kullback's test tends to be too liberal; therefore, we present Jennrich's derived test, which, like Kullback's test, is also very general (and in fact, includes as a special case testing if a correlation matrix R is equal to a population matrix P, as shown in the previous section). This test can be used to test special-case hypotheses such as the equality of two correlation matrices and the equality of more than two correlation matrices (the most general case). We split our treatment of Jennrich's test into these two sections to facilitate providing examples in each of these cases and to show how the test statistic simplifies in special cases; in addition, we cover the special case of comparing more than two independent correlations.

Note that these tests are designed to compare correlation matrices from independent groups. If a researcher has two correlation matrices measured on the same group—say, a correlation matrix computed on variables at Time 1, and a correlation matrix computed for variables at Time 2, for which the participants at Time 1 and Time 2 are the same—these two correlation matrices are, in fact, part of a *larger* correlation matrix, and the researcher may use methods presented in the above section to compare correlations within this larger correlation matrix.

Testing Equality of Two Correlation Matrices From Independent Groups

Null Hypothesis

$H_0 : P_1 = P_2$ (two correlation matrices from independent samples are equal in the population)

Test Statistic

$$\chi^2 = \frac{1}{2}\text{tr}\left(Z^2\right) - \text{diag}'(Z)S^{-1}\text{diag}(Z)$$

In this test statistic, $Z = \sqrt{\dfrac{N_1 N_2}{N_1 + N_2}}\left(\bar{R}\right)^{-1}\left(R_1 - R_2\right)$; $S = \bar{R} \odot \left(\bar{R}\right)^{-1} + I$;

diag() are the diagonal elements of a matrix (organized into a column vector); tr() is the trace, calculated by summing the diagonal elements; $\bar{R}$ is the weighted average of the two correlation matrices, with each correlation matrix weighted by its sample size; N_1 and N_2 are the sample sizes of the correlation matrices, respectively; $\left(\bar{R}\right)^{-1}$ is the inverse matrix of $\bar{R}$; $\odot$ denotes multiplication of corresponding elements (vs. standard matrix multiplication, which does not involve simple multiplication of corresponding elements); and I is the identity matrix of order p, which has 1s along the diagonal and 0s everywhere else. The test statistic is asymptotically distributed as χ^2 with $df = \dfrac{p(p-1)}{2}$.

Example. For this example, we will use the boys' and girls' intelligence matrices, as presented in Tables 1.1 and 1.2. Letting R_1 equal the correlation matrix in Table 1.1 (boys), and R_2 equal the correlation matrix in Table 1.2 (girls), and letting $N_1 = 391$ and $N_2 = 495$ (the lowest reported sample sizes for each group, respectively),

$$\bar{R} = \frac{N_1}{N_1 + N_2}R_1 + \frac{N_2}{N_1 + N_2}R_2 = \frac{391}{886}R_1 + \frac{495}{886}R_2$$

$$\bar{R} = \begin{bmatrix}
1.00 & .639 & .636 & .687 & .668 & .618 & .622 & .631 & .626 & .536 \\
 & 1.00 & .690 & .689 & .708 & .708 & .686 & .601 & .606 & .592 \\
 & & 1.00 & .784 & .776 & .717 & .739 & .721 & .699 & .594 \\
 & & & 1.00 & .876 & .791 & .772 & .796 & .804 & .750 \\
 & & & & 1.00 & .846 & .818 & .781 & .794 & .766 \\
 & & & & & 1.00 & .850 & .759 & .770 & .781 \\
 & & & & & & 1.00 & .823 & .783 & .750 \\
 & & & & & & & 1.00 & .891 & .786 \\
 & & & & & & & & 1.00 & .861 \\
 & & & & & & & & & 1.00
\end{bmatrix}$$

Using $\bar{R}$ to calculate $\left(\bar{R}\right)^{-1}$ with software:

$$\left(\bar{R}\right)^{-1} = \begin{bmatrix} 2.222 & -.601 & -.129 & -.561 & -.241 & -.013 & -.026 & -.271 & -.385 & .421 \\ & 2.598 & -.658 & -.198 & -.236 & -.585 & -.394 & .383 & .255 & -.264 \\ & & 3.441 & -1.04 & -.687 & -.116 & -.527 & -.482 & -.440 & .966 \\ & & & 5.757 & -2.55 & -.217 & .433 & -.686 & -.595 & -.435 \\ & & & & 6.467 & -1.54 & -.771 & .216 & -.273 & -.519 \\ & & & & & 5.373 & -2.07 & .332 & .047 & -1.18 \\ & & & & & & 5.316 & -2.00 & .293 & -.229 \\ & & & & & & & 6.360 & -3.66 & -.053 \\ & & & & & & & & 7.713 & -3.05 \\ & & & & & & & & & 4.847 \end{bmatrix}$$

Finally, calculating Z and s^{-1}:

$$Z = \sqrt{\frac{N_1 N_2}{N_1 + N_2}} \left(\bar{R}\right)^{-1} \left(R_1 - R_2\right)$$

$$Z = \begin{bmatrix} 1.17 & -2.56 & -1.15 & -1.14 & -1.11 & -1.14 & -1.04 & -0.25 & -0.45 & 0.20 \\ -1.77 & 0.36 & 3.62 & 1.36 & -1.01 & -0.90 & 0.35 & -0.68 & -0.76 & -1.47 \\ 1.60 & 4.71 & -0.86 & 0.82 & 0.57 & -0.96 & 1.91 & -0.71 & 0.50 & 1.25 \\ -0.05 & 3.42 & 1.23 & 0.23 & -0.24 & -0.38 & -2.04 & 0.80 & 0.80 & 0.76 \\ -1.66 & -4.66 & -1.38 & -0.18 & 1.60 & 0.67 & -3.33 & -1.83 & 0.19 & -0.38 \\ -0.82 & -2.14 & -3.83 & -0.37 & 1.31 & 1.23 & 1.49 & 1.53 & -0.87 & -1.09 \\ -0.17 & 1.45 & 2.79 & -2.26 & -2.63 & 0.01 & -0.52 & 2.14 & 2.87 & 1.09 \\ -0.58 & -1.98 & -3.29 & -0.27 & -0.90 & 2.20 & 1.40 & 0.45 & -2.98 & -0.25 \\ 0.45 & 1.51 & 3.33 & 1.72 & 2.70 & 0.33 & 2.41 & -1.35 & 1.87 & -1.87 \\ 1.46 & -0.39 & -0.11 & -0.00 & -1.00 & -1.79 & -0.66 & -0.38 & -1.40 & 1.57 \end{bmatrix}$$

$$S^{-1} = \left[\bar{R} \odot \left(\bar{R}\right)^{-1} + I \right]^{-1}$$

$$S^{-1} = \begin{bmatrix} .323 & .040 & .023 & .032 & .022 & .011 & .012 & .020 & .018 & .000 \\ & .296 & .040 & .022 & .026 & .035 & .027 & .000 & .002 & .013 \\ & & .251 & .048 & .040 & .022 & .032 & .029 & .021 & -.005 \\ & & & .185 & .068 & .029 & .017 & .033 & .035 & .031 \\ & & & & .173 & .054 & .038 & .021 & .026 & .033 \\ & & & & & .195 & .067 & .019 & .022 & .047 \\ & & & & & & .197 & .055 & .027 & .029 \\ & & & & & & & .190 & .087 & .045 \\ & & & & & & & & .176 & .085 \\ & & & & & & & & & .223 \end{bmatrix}$$

$$\chi^2 = \frac{1}{2}\text{tr}\left(Z^2\right) - \text{diag}'\left(Z\right)S^{-1}\text{diag}\left(Z\right) = \frac{1}{2}\text{tr}\left(188.25\right) - 4.13 = 89.99$$

The $\alpha = .05$ critical value for a χ^2 distribution with $df = \dfrac{10(10-1)}{2} = 45$ is 61.66. Because $\chi^2 > 61.66$, we reject the null hypothesis and conclude that the correlation matrices for boys' and girls' intelligence differ in the population. This is different from what we might have expected based on casual observation of the correlation matrices because they appear similar, but given the large sample sizes, the test is powerful enough to reject equality even for these relatively small differences.

Testing Equality of Several Correlation Matrices

In this section, we present the most general form of the Jennrich (1970) test, which may be used for any number of m independent correlation matrices each of size (order) p.

Null Hypothesis

$H_0 : P_1 = P_2 = ... = P_m$ (m correlation matrices from independent samples are equal in the population)

Test Statistic

$$\chi^2 = \sum_{i=1}^{m} \frac{1}{2}\text{tr}\left(Z_i^2\right) - \text{diag}'\left(Z_i\right)S^{-1}\text{diag}\left(Z_i\right)$$

In this test statistic, $Z_i = \sqrt{N_i}\left(\bar{R}\right)^{-1}\left(R_i - \bar{R}\right)$; $S = \bar{R} \odot \left(\bar{R}\right)^{-1} + I$; diag() are the diagonal elements of a matrix (organized into a column vector); tr()

is the trace as calculated by summing the diagonal elements of the matrix; $\bar{R}$ is the weighted average of the two correlation matrices, with each correlation matrix weighted by its sample size; N_1 and N_2 are the sample sizes of the correlation matrices, respectively; $\left(\bar{R}\right)^{-1}$ is the inverse matrix of $\bar{R}$; $\odot$ denotes multiplication of corresponding elements; and I is the identity matrix of order p, which has 1s along the diagonal and 0s everywhere else. The test statistic is asymptotically distributed as χ^2 with $df = \dfrac{p(p-1)(m-1)}{2}$.

Example. For this example, we will use the racial composition of sports teams data to test if the correlation matrices for MLB (R_1), NFL (R_2), and the NBA (R_3) are the same. Again, we assume that $N = 26$, which is likely to be a slight underestimate.

$$R_1 = \begin{bmatrix} 1.0 & .09 & -.18 & -.23 \\ & 1.0 & .11 & .04 \\ & & 1.0 & .96 \\ & & & 1.0 \end{bmatrix}, \quad R_2 = \begin{bmatrix} 1.0 & .13 & -.05 & -.10 \\ & 1.0 & .36 & .30 \\ & & 1.0 & .99 \\ & & & 1.0 \end{bmatrix},$$

$$R_3 = \begin{bmatrix} 1.0 & .41 & -.06 & -.06 \\ & 1.0 & .37 & .29 \\ & & 1.0 & .99 \\ & & & 1.0 \end{bmatrix}$$

$$\bar{R} = \frac{N_1}{\sum N_i}R_1 + \frac{N_2}{\sum N_i}R_2 + \frac{N_3}{\sum N_i}R_3 = \begin{bmatrix} 1.0 & .21 & -.10 & -.13 \\ & 1.0 & .28 & .21 \\ & & 1.0 & .98 \\ & & & 1.0 \end{bmatrix}$$

$$\left(\bar{R}\right)^{-1} = \begin{bmatrix} 1.09 & -0.24 & -0.40 & 0.58 \\ & 1.28 & -2.21 & 1.86 \\ & & 29.7 & -28.7 \\ & & & 28.8 \end{bmatrix};$$

$$S = \bar{R}\left(\bar{R}\right)^{-1} + I = \begin{bmatrix} 2.09 & -0.05 & 0.04 & -0.08 \\ & 2.28 & -0.62 & 0.39 \\ & & 30.7 & -28.1 \\ & & & 29.8 \end{bmatrix}$$

$$Z_1 = \begin{bmatrix} 0.0 & -0.8 & -0.3 & -0.3 \\ -0.8 & 0.4 & -1.2 & -0.8 \\ 3.4 & -0.6 & 5.0 & -0.9 \\ -3.6 & -0.5 & -4.8 & 1.0 \end{bmatrix}, \quad Z_2 = \begin{bmatrix} 0.1 & -0.3 & 0.2 & 0.0 \\ -0.8 & 0.1 & 0.6 & 0.4 \\ 3.6 & -0.9 & -2.5 & 0.4 \\ -3.2 & 1.3 & 2.4 & -0.5 \end{bmatrix},$$

$$Z_3 = \begin{bmatrix} -0.1 & 1.2 & 0.1 & 0.3 \\ 1.6 & -0.5 & 0.6 & 0.3 \\ -6.9 & 1.5 & -2.5 & 0.5 \\ 6.8 & -0.8 & 2.4 & -0.5 \end{bmatrix}.$$

$$\chi^2 = [(0.5)(38.7) - 8.8] + [(0.5)(10.2) - 2.1] + [(0.5)(16.3) - 2.4] = 10.5$$
$$+ 3.0 + 5.7 = 19.2$$

The $\alpha = .05$ critical value for a χ^2 distribution with $df = \dfrac{(4)(4-1)(3-1)}{2} = 12$ is 21.03. Because $\chi^2 < 21.03$, we fail to reject the null hypothesis and cannot conclude that the correlation matrices for the MLB, NFL, and NBA are different from each other in the population.

Testing Equality of Several Correlations From Independent Samples

Suppose we have m independent samples and use each sample to calculate a correlation between two variables of interest. For example, suppose we are interested in the bivariate correlation between proximity to work and gross income in m U.S. cities. We can test the equality of $r_1, r_2, \ldots r_m$ by treating each correlation as the sole element of a 2×2 correlation matrix. This test cannot be performed with the Jennrich (1970) statistic; when $p = 2$, the Jennrich (1970) test statistic will always produce a value of $\chi^2 = 0$. As such, we suggest Kullback's (1967) H statistic, which is a simplified formula relative to Kullback's most general H statistic.

Null Hypothesis

$H_0 : \rho_1 = \rho_2 = \ldots = \rho_m$ (m correlations from independent samples are equal in the population)

Test Statistic

$$H = \sum_{i=1}^{m} (N_i - 1) \left[\ln \frac{1 - \bar{r}^2}{1 - r_i^2} \right]$$

where $\bar{r}$ is the weighted average of the m correlations, weighted by the sample sizes (-1). H is asymptotically distributed as χ^2 with $df = m - 1$.

Example. Suppose we have four independent correlation coefficients: $r_1 = 0.21$, $r_2 = 0.25$, $r_3 = 0.18$, and $r_4 = 0.20$, each computed on a sample of $N_1 = N_2 = N_3 = N_4 = 16$ people. Then,

$$\bar{r} = \frac{N_1 - 1}{\sum N_i - 4} r_1 + \frac{N_2 - 1}{\sum N_i - 4} r_2 + \frac{N_3 - 1}{\sum N_i - 4} r_3 + \frac{N_4 - 1}{\sum N_i - 4} r_4$$

$$= \frac{15}{60}(0.21 + 0.25 + 0.18 + 0.20) = 0.21$$

$$H = \sum_{i=1}^{4} (N_i - 1)\left| \ln \frac{1 - \bar{r}^2}{1 - r_i^2} \right|$$

$$= 15\left[\ln \frac{1 - (0.21)^2}{1 - (0.21)^2} \right] + 15\left[\ln \frac{1 - (0.21)^2}{1 - (0.25)^2} \right]$$

$$+ 15\left[\ln \frac{1 - (0.21)^2}{1 - (0.18)^2} \right] + 15\left[\ln \frac{1 - (0.21)^2}{1 - (0.20)^2} \right]$$

$= 15[0] + 15[0.0194] + 15[-0.0122] + 15[-0.00428] = 0.0449.$

The $\alpha = .05$ critical value for a χ^2 distribution with $df = 4 - 1 = 3$ is 7.815. Because $H \leq 7.815$, we fail to reject the null hypothesis and, therefore, cannot conclude that these four correlations are different in the population.

Testing for Linear Trend of Eigenvalues

The scree test is a common (but imprecise) method for determining the number of factors in exploratory factor analysis (EFA; see Chapter 4). The scree test involves inspecting the eigenvalues of the modified correlation matrix (displayed in a "scree plot") to determine where a noticeable "drop" in the eigenvalues occurs. The eigenvalues after the drop usually follow a linear trend, whereas the eigenvalues before the drop are less likely to follow a linear trend. The smallest eigenvalues that tend to follow a linear trend are generally associated with trivial or nuisance dimensions, as opposed to the dominant or substantive dimensions that are associated with the larger eigenvalues.

Because scrutinizing a scree plot for the drop in eigenvalues may be subjective, researchers have instead developed tests for linearity of the smallest eigenvalues of a matrix. Anderson (1963) developed a test for assessing the linearity of the last q eigenvalues of a covariance matrix, proposing that the same test may serve as a test for eigenvalues of a correlation matrix. Anderson noted that this test is likely to be conservative (i.e., more likely to support linearity in the eigenvalues when they actually deviate from linearity in the population) for correlation matrices, and Bentler and Yuan (1998) showed empirically that Anderson's statistic is conservative across a range of conditions. However, we present Anderson's statistic (called T) rather than the more complex statistic developed by Bentler and Yuan because the Bentler–Yuan statistic requires iteration to calculate and also has low power for sample sizes often seen in practice.

Null Hypothesis

$H_0 : \lambda^*_{p-q+j} = a - bj$, $j = 1, 2, \ldots, q$ (there is a linear trend in the last q eigenvalues of P)

Test Statistic

$$T = -\left(N - 1 - p + q - \frac{2q^2 + q + 2}{6q}\right) \ln \frac{\prod_q \lambda}{\bar{\lambda}^q}$$

where $\prod_q \lambda$ is the product of the last q eigenvalues, and $\bar{\lambda}$ is the average of the last q eigenvalues. T is distributed as χ^2 with $df = \frac{(q+2)(q-1)}{2}$.
Example. For the NLSY fertility data set (smoothed, as shown in the lower triangular of Table 2.1), the five eigenvalues of the correlation matrix are

$$\lambda = (2.15, 1.57, 0.65, 0.49, 0.14)$$

Suppose we wish to test whether the last three eigenvalues follow a linear trend. Then, $q = 3$, $p = 5$, $N = 7,000$.

$$\bar{\lambda} = \frac{0.65 + 0.49 + 0.14}{3} = 0.427$$

$$\prod_q \lambda = 0.65 \times 0.49 \times 0.14 = .0451$$

$$T_A = -\left(7000 - 1 - 5 + 3 - \frac{2(3)^2 + 3 + 2}{6(3)}\right) \ln \frac{0.0451}{0.427^3} = 3819.79$$

The critical value for a χ^2 distribution with $df = \dfrac{(3+2)(3-1)}{2} = 5$ is 11.07, so we would reject the null hypothesis and conclude that the last three eigenvalues for this correlation matrix deviate from a linear trend. (Note that the large sample size provides very high power to detect deviations from linearity in this sample.)

Summary

The starting point—though often far from the finishing point—for statistically evaluating correlations and correlation matrices is to conduct NHST in relation to relevant null hypotheses. In Chapter 4, we expand the types of models that can be used in relation to correlation matrices and treat factor analysis, PCA, and SEM. In the current chapter, we treated more basic statistical issues.

The primary goal of the current chapter was to organize, and then to illustrate, how a researcher can test hypotheses related to correlation matrices within an NHST framework. Substantial mathematical detail exists in the background, associated with the development of sampling distributions for the various hypothesis tests that we have developed in this chapter. However, all NHST methods are united by a common logic, which necessarily unites all of the hypothesis testing approaches defined in the current chapter. We have used the data examples from the first chapter to illustrate these methods and have provided cross-disciplinary examples in the use of those data sets. Specifically, we used the intelligence data, the sports team data, the trust-in-government data, and the NLSY fertility data from the correlation matrices in Chapter 1 for illustration. In the next chapter, we develop the modeling methods that provide the "next step"—methods that further reveal the substantial information that is captured within a correlation matrix.

Chapter 4

METHODS FOR CORRELATION/COVARIANCE MATRICES AS THE INPUT DATA

In this chapter, we treat the correlation matrix not as a descriptive tool of the raw data, but rather as the input data to be directly analyzed. We introduce statistical methods that are commonly employed in the social sciences for correlation matrices. These methods often fall into the category of statistical modeling, a broad methodological perspective that includes hypothesis testing as a special case (e.g., Rodgers, 2010). Many modern statistical modeling methods exist that treat correlations as input data, including path analysis, structural equation modeling (SEM), factor analysis, components analysis, multilevel modeling, and latent growth curve modeling, among others.

In this introduction to this chapter, we present some brief comments about models and about correlation matrices. However, we leave in-depth treatment for specialized sources specifically dedicated to these topics. When a researcher has collected data and computed a correlation (or covariance) matrix based on the data, it is typical to assume that some model—some definable, simplifying process—generated the patterns in the correlation matrix. As researchers, we'd like to identify one or more potential models that may have generated the empirical correlational patterns.

Models are, by definition, simplifications of the complex reality that researchers study. In addition to simplifying reality, though, models are also expected to provide a match to that reality (what researchers developing models refer to as the "goodness of fit" of the model). For example, a researcher studying adolescent sexual behavior can develop models from a number of different perspectives (see Rodgers, 2000, for treatment that is expanded well beyond the current discussion). One way to simplify the complex reality, and also to partially match that reality, is to model adolescent sexual behavior biologically, as it is stimulated physiologically by hormonal processes that develop during puberty. Another way to model adolescent sexual behavior is psychologically, by studying the social influence from an adolescent friendship network. Another way is sociologically, by modeling adolescent sexual behavior through the mitigation of such behavior by religious norms. Each model—each simplification—would generate correlations in a particular measurement domain that could be investigated. The physiological domain would be reflected in correlations between height and weight growth processes, onset of menstrual cycles, sexual motivation, and sexual behavior—all potentially influenced by hormonal development. The psychological domain could be measured using indicators of social connectedness, sexual content in peer conversations, and sexual behavior. The sociological domain could be studied using a correlation matrix reflecting relationships among parental spirituality, attendance at religious services, conservative versus progressive religious views, and adolescent sexual behavior.

The methods in this chapter use matrix algebra approaches to decompose the patterns in correlation matrices. Following, those decompositions of the correlational patterns are interpreted substantively. The goal of each decomposition is to attempt to identify one or more potential models that may have generated the correlational patterns.

Factor Analysis

Summary

A consistent theme throughout this book is that correlation matrices are much more interesting and valuable than the combined pairwise correlation coefficients that define them. Correlation matrices allow a larger scope and more efficiency with which to view the relationships among variables; they are truly multivariate, whereas correlation coefficients are by definition bivariate. Emerging from this principle is the modeling approach called factor analysis, which is a method for analyzing correlation matrices (literally, "breaking down correlation matrices") to identify underlying constructs (factors) that cause the variables to correlate the way they do. For example, a researcher may have several items measuring personality traits to which participants in a study may respond, such as participants' agreement with the statements "I like to be on time," "I am flexible and accommodating when plans change," or "I like to meet new people." These items likely correlate due to underlying personality factors (e.g., contentiousness, extraversion) that inform participants' responses to items. The measured items are called "manifest variables" because they are the measured manifestations of the underlying personality traits, whereas the personality traits would be called "latent variables" or "factors" because they are unmeasured (and often unmeasurable), but investigated through multiple manifest variables. Factor analysis is ultimately the study of latent variables, though the researcher only gets to directly observe the manifest variables.

There are two branches of factor analysis—exploratory and confirmatory—that both operate toward the goal of identifying and studying latent variables, or factors, that underlie the correlation matrix. Exploratory factor analysis (EFA) is concerned with deciding on (or "discovering," depending on the prevailing epistemological view) the underlying factors that cause variables to correlate the so-called factor structure. EFA is usually conducted when developing a new scale for a construct of interest; through EFA, a researcher can investigate subscales of the overall scale, or decide which items should be added, deleted, or altered. Confirmatory factor analysis (CFA), alternatively, tests a theorized factor structure using the empirical information in a set of items and the correlations among them. A researcher who believes their measure (developed, perhaps, through EFA) can appropriately assess the constructs of interest may test their measure's effectiveness through CFA. If the CFA model fits well, this excellent fit may indicate that the hypothesized factor structure has merit; if the CFA model does not fit well, the researcher may have to go back to EFA to reinvestigate the factor structure. Both EFA and CFA allow underlying factors themselves to be uncorrelated (orthogonal) or correlated (oblique). Importantly, EFA and CFA should not be applied to the same correlation matrix, as a prior EFA may improperly inform the theory being tested on the same correlation matrix in the CFA, increasing Type II errors.

Factor analysis is closely related to PCA. Both were developed as tools to determine underlying structure of manifest (i.e., measured) variables, as captured in a correlation matrix, but they developed from different disciplines and with different goals. In factor analysis, the latent variables (factors) are often of theoretical significance or of primary interest to the researcher. In PCA, the underlying variables (called "components") are often unnamed or not of theoretical interest—they primarily serve to reduce the dimensionality of the correlation matrix, that is, as a data reduction method. PCA is carried out directly on the correlation matrix; the largest component (the principal component) corresponds to the direction of the first eigenvector and accounts for the direction of maximal variance (measured by the first eigenvalue) among the variables. It is in this respect that factor analysis and PCA differ statistically. For factor analysis, the correlation matrix is modified in such a way that measurement error of the variables is accounted for by an adjustment in the diagonal elements of the correlation matrix. Specifically, the estimates of the common variance shared by a given variable with the rest of the variables are inserted into the diagonal elements. These estimates are called "commonalities." In PCA, no adjustment is made to the diagonals, which are all defined as equal to 1 (which is by definition the correlation of a variable with itself, and would naturally be defined as the diagonal element). These methods reflect the different goals of factor analysis and PCA; the latter is concerned with maximizing the amount of variance that is explained by the model, whereas the former has the goal of reproducing the correlation matrix.

Using software, we can conduct either EFA or CFA. The required input to these models is either a correlation or covariance matrix, or the raw data from which those matrices can be computed. Both EFA and CFA results include estimates of path coefficients, which are interpreted as regression coefficients from regressing a manifest variable (the dependent variable) on the latent variable (the predictor), and correlations between latent variables (if allowed by the researcher). In addition, CFA results include model goodness-of-fit measures (of which there are many), meant to indicate how well the proposed CFA model fits the data.

Example

Consider the correlation matrix between nine survey items measured on 6,000+ respondents in the NLSY, presented in Table 1.7. Cursory inspection of the correlations indicates that items may be grouped into three subsets: (1) Four of the items appear to represent self-esteem or optimism, (2) three of the items appear to represent willingness to take risk for positive change, and (3) the final two items appear to represent openness to new people and experiences. Correlations within each subset are of mid- to high magnitude, with self-esteem items more strongly correlated than risk-taking items, which are more strongly correlated than the two openness items; correlations across the subsets are much smaller. One may therefore posit that these nine items may be well represented by three latent factors, one associated with each subset of items—(1) Self-Esteem, (2) Positive Risk Taking, and (3) Openness—with each item associated with exactly one latent factor and with the latent factors allowed to correlate with each other.

A CFA was conducted on the overall 9 × 9 correlation matrix to estimate the path coefficients and to calculate the correlations among the three latent factors. The four (apparent) self-esteem items have standardized path coefficient estimates of .81, .85, .71, and .62 as predicted by the factor we interpret as Self-Esteem; the three (apparent) risk-taking items have standardized path coefficient estimates of .69, .54, and .73 as predicted by the factor we interpret as Positive Risk Taking; and the two (apparent) openness items have standardized path coefficient estimates of .56 and .52 as predicted by the factor we interpret as Openness. These path coefficients may be interpreted as standardized regression coefficients (regression coefficients estimated from standardized variables) or as correlations between each item and its corresponding latent variable. Note that the path coefficients are higher than the raw correlations because the factors are assumed within the CFA model to be measurement-error free, unlike their corresponding manifest variables, and measurement error attenuates observed correlations. Note also that the path coefficients for Self-Esteem are higher than the path coefficients for Positive Risk Taking, which are higher than the path coefficients for Openness. This pattern makes sense given the measurement of the items in the NLSY: The self-esteem items come from the psychometrically validated Rosenberg Self Esteem Scale; the risk-taking items come from a larger set of items measuring risk taking and are worded similarly; and the openness items are also from a larger set of items on personality but are not meant to measure the specific underlying construct "Openness." These path coefficient estimates, although larger than the interitem correlations, are not particularly encouraging by most standards for path coefficients; to more fully measure latent Self-Esteem, Positive Risk Taking, and Openness, we'd ideally want to add more items or change the wording of some items to improve their measurement properties.

We also estimated correlations between the three latent factors themselves, thereby reducing the 9 × 9 correlation matrix of observed variables to a 3 × 3 correlation matrix of latent factors. Self-Esteem and Openness correlated the highest ($r = .28$), followed by Positive Risk Taking and Openness ($r = .19$), and Self-Esteem and Positive Risk Taking ($r = .09$). As above, the correlations between the factors are higher than the average interitem correlations between the observed items because the latent factors are assumed to be measurement-error free.

Although we omit technical details about model syntax, estimation, and fit, it is worth noting that this CFA model fits acceptably, though not superbly, as measured by a number of fit indices. A researcher with these CFA results may be encouraged to explore model misfit—perhaps by changing items to more properly reflect the underlying factors or by adding paths in the model that may be theoretically driven (e.g., correlated residuals could be assumed between items that have particularly similar wording or meaning, such as "I have a positive attitude with myself and others" and "I am satisfied with myself").

Resources for Software and Additional Readings

A number of excellent texts exist that cover factor analysis in detail. Two books in the QASS series by authors Kim and Mueller (1978a, 1978b) describe factor analysis conceptually, as well as the underlying mathematical equations. Further distinctions between EFA and CFA, as well as practical data issues, are also discussed. The second edition of *Confirmatory Factor Analysis for Applied Research* (Brown, 2015) may also be of interest to researchers for an updated, practical discussion of issues common to CFA. For researchers who are interested in assessment, Bandalos (2017) offers a practical guide to measurement issues involved in factor analysis and related methods that integrates classical test theory and more modern approaches.

Both EFA and CFA may be conducted in common statistical software used in the social sciences, including SAS, R, Stata, and Mplus. Software programs differ, however, in their offerings of advanced EFA and CFA techniques. Resources for these software packages are constantly developing; we recommend that the applied researcher interested in conducting factor analysis should consult the user manual or program documentation for their preferred statistical package for syntax; we find that most examples included in these resources are excellent for demonstrating simple factor analyses and describing options available in the statistical package for advanced analyses.

Structural Equation Modeling

Summary

SEM, estimated from correlation or covariance matrices, combines factor analysis and regression to model how variables relate to each other, either through directed (causal) relationships or undirected (correlational) relationships. One part of an SEM defines how manifest (measured) variables are used to define a latent variable (a construct); this part of the model is referred to as the "measurement model." The other part of an SEM defines how the latent variables (the constructs) relate to one another (the "structural model"). The measurement model is closely related to the measurement features of factor analysis, with the latent variable serving as the factor, defined by the manifest variables. The structural model can be considered to be the combination of several (potentially overlapping) regression models, with one or more latent variables predicted from multiple latent variables. Thus, SEM is versatile and includes as special cases other modeling frameworks, some of which are beyond the scope of the current treatment; besides factor analysis, SEM can be used to estimate models related to path analysis, the general linear model, multilevel modeling, latent growth curve modeling, mixture modeling, and autoregressive models. Each variable in an SEM can include explicitly defined measurement error, and measurement errors can be correlated (or not) between variables. SEM can test hypotheses such as how two variables (manifest or latent) relate to each other, how the overall model fits the data, where within the model there is notable lack of fit, or which model of several competing models fits the data best. It is important to emphasize, within the context of this book, that the SEM methodology begins with a correlation or covariance matrix. In fact, the early versions of SEM were referred to as "analysis of covariance structures" (e.g., Bentler & Bonett, 1980; Jöreskog, 1970).

A feature that makes SEM particularly popular is that a hypothesized SEM can be represented graphically, rather than algebraically, in a path diagram. For example, consider the path diagram in Figure 4.1, which shows a theoretical model for how several variables measuring fertility expectations and fertility outcomes (from Table 1.6) may be related. The rectangles represent the measured variables, and the covariance or correlation matrix underlying the variables in the rectangles would serve as the input matrix to the SEM analysis. The single-headed arrows represent the causal or predictive relationships (paths) of one variable to the other. From the paths, we can see that fertility expectations are theorized to predict fertility outcomes, and past fertility outcomes are theorized to predict future fertility outcomes. This path diagram corresponds to a path analysis, which is a special case of SEM akin to simultaneous regression. In path analysis, the manifest variables are directly modeled, without reference to latent variables. The only latent variables (represented with ovals/circles rather than rectangles) in this particular path analysis are the residuals. Variances and covariances are represented with curved, double-headed arrows.

Figure 4.1 Theorized Path Diagram for Fertility Expectations and Outcomes

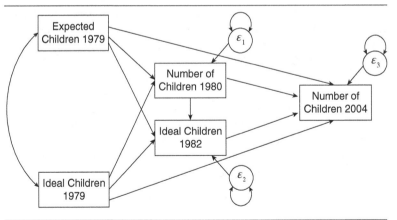

Note: Directed arrows indicate directions of cause or prediction; for example, Ideal Number of Children in 1979 predicts Ideal Number of Children in 1982. Double-headed arrows between variables indicate assumed covariance/correlation between variables. Double-headed arrows are also used on the residuals of predicted variables (e.g., the ϵs) to show residual variance.

In the SEM framework, CFA can also be graphically demonstrated in a straightforward way. We demonstrate a CFA in an example earlier in this chapter, in which an SEM is fit to the correlation matrix from Table 1.7.

Besides the inherent interpretability of path diagrams, SEM has also gained ground in popularity for its ability to test specialty hypotheses that may be convoluted, or impossible, to test in other modeling frameworks. For example, group differences can be tested naturally in SEM by constraining some, or all, paths to be equal between two or more groups. This is a common approach to testing measurement invariance across groups—for instance, comparing and testing a measure's factor structure for men versus women, or typical versus atypical subpopulations. SEM is amenable to observational research and experimental research alike. Like CFA, SEM output provides path coefficient estimates and multiple model fit indices, all obtained through analysis of correlation/covariance matrices.

SEM, though flexible, is not a panacea for all analytic circumstances. SEM is a large-sample technique; common sample size recommendations are in the hundreds, depending on the complexity of the model tested. These sample sizes are necessary to ensure the stability of the correlation/ covariance matrices that are typically used as input to SEM analysis. In addition, the statistical theory underlying the usual SEM estimation routines—often based on maximum likelihood estimation—requires large sample sizes. Model identification and estimation (which we conveniently omit in this coverage) are nontrivial issues in conducting SEM; a theorized SEM model may not be "identified," meaning no unique set of path estimates exists for the model, even if the theorized model closely reflects reality. To determine whether a model is identified requires, among other things, counting the number of correlations or covariances and ensuring that count is greater than the number of parameters to be estimated in the model. We illustrate this counting process in the two examples that follow. We also note that even with an identified model and a large enough sample, complicated models fail to properly converge on a solution in a significant percentage of real applications, and sometimes there are multiple SEM models that are approximately equivalent but that imply radically different interpretations of relationships among variables.

Examples

Fertility Expectations and Outcomes Example. Using the correlation matrix presented in the lower triangular half of Table 2.1, which represents a smoothed version of the correlation matrix of fertility outcomes and expectations first presented in Table 1.6, the relationships in the path

55

diagram presented in Figure 4.1 were estimated using path analysis. In this estimation process, the path coefficients are estimated to optimize a specified objective function (usually path analysis/SEM uses maximum likelihood estimation, which is what we use here) that computes the "best" path coefficients to reproduce the covariance/correlation matrix. The path coefficient estimates are displayed next to their respective paths in Figure 4.2 and can be interpreted as standardized regression coefficients. This path diagram shows how intermediate fertility outcomes and expectations may mediate the relationship between early fertility outcomes, when respondents were 14 to 22 years old, and later fertility outcomes, when respondents are 39 to 47 years old, and fertility is mostly completed. Surprisingly, ideal and expected number of children reported in 1979 do not operate to predict future fertility expectations and outcomes in similar ways; expected number of children was a better predictor of number of children in 2004, controlling for other predictors in the model, and ideal number of children reported in 1982 was not as strong a predictor of number of children in 2004 as expected number of children in 1979. Less surprising is that the number of children born by 1980 was the best predictor of the number of children born by 2004; earlier behavior often predicts later behavioral outcomes better than attitudes and expectations.

Figure 4.2 Path Analysis Using the Correlations Among Fertility Outcomes and Expectations Presented in Table 2.1 (Lower Triangular)

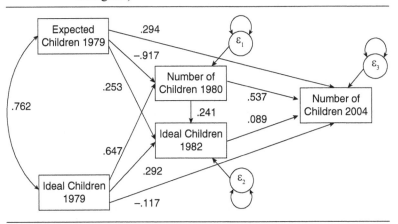

Note: The path coefficients are interpreted as standardized regression coefficients.

This example brings up a couple of interesting issues. First, in SEM parlance, this path analysis is just-identified—the number of path coefficients estimated is equal to the number of correlations in the correlation matrix (a total of 10). In the language of statistics, the number of degrees of freedom that were originally available ($df = 10$) were all used in estimating the model; none were left over to estimate the residual variance that is used to evaluate the quality of fit of the model (see Rodgers, 2019, for in-depth development of the degrees of freedom concept in the context of statistical pedagogy). In an SEM sense, this model fits perfectly, meaning the observed correlations (as presented in Table 2.1) are perfectly reproduced from those that are implied by the model's path estimates. This "perfect fit" implies that all of the SEM fit indices will indicate that this model perfectly explains the data. However, this does not mean this model is "correct," or even good/excellent, because the perfect fit is an artifact of the just-identified nature of the model-fitting exercise. Another path diagram may be drawn that is also just-identified, and therefore also has perfect fit, which indicates fundamentally different relationships between the variables. An SEM that closely approximates reality should fit the data well, but good model fit does not necessarily mean the model accurately represents reality. Ideally, there should be many more data points (in a path analysis or SEM, that means "many more correlations or covariances") than estimated parameters (path coefficients). As a result, some of the degrees of freedom are used to estimate the model, and the "extra" left-over data points represent degrees of freedom that can be used to test the fit of the model to the data (which cannot occur in the fitting exercise presented in Figure 4.2). Viewing degrees of freedom as a "data accounting process" is discussed in detail in Rodgers (2019).

Second, because the correlation matrix used was smoothed from a non-PSD matrix to a PD matrix, this correlation matrix has a positive eigenvalue that is near zero. Very small positive eigenvalues may cause some instability in the estimates; for example, the path coefficient between expected number of children in 1979 and number of children in 1980 is a whopping $-.917$, which is a large standardized value and is likely far from its true population value. The bias in this estimate, due to the ill-conditioned nature of the correlation matrix, is likely to bias other path estimates as well.

Self-Esteem, Openness, and Positive Risk Taking Mediation Example. The next example shows an SEM that includes both latent and manifest variables. In this analysis, we used the correlation matrix from Table 1.7 regarding survey items for Self-Esteem, Openness, and Positive Risk Taking to fit a CFA with three latent factors. We continue that example by replacing the correlational paths between latent variables with directed paths between the latent variables—the so-called structural paths, as they denote the structure of the relationships among the latent variables. Specifically, we allow Openness to completely mediate the relationship between Self-Esteem and Positive Risk Taking, with no direct relationship between Self-Esteem and Positive Risk Taking.

This SEM example includes latent variables, represented in ovals. The path coefficient estimates from latent variables to their manifest variables are similar to the estimates obtained in the corresponding CFA, but they are not identical. The structural path estimates are also similar to their corresponding correlation estimates in the CFA; this is not generally the case, but in this example, the correlation between Self-Esteem and Positive Risk Taking was so small ($r = .09$) that its omission did not greatly alter the other relationship estimates between the latent variables. Because we fit a model estimating 11 parameters to a correlation matrix with 36 correlations, we had 25 degrees of freedom remaining after we estimated the path coefficients in Figure 4.3. We used the information associated with these 25 residual degrees of freedom to evaluate the fit of the model that was obtained. Like the CFA, this model fit adequately, but not superbly.

Resources for Software and Additional Readings

We recommend several resources for readers interested in introductory, advanced, or practical SEM knowledge alike. *Principles and Practice of Structural Equation Modeling* (Kline, 2015) is structured to be

Figure 4.3 Structural Equation Model Using Data From Table 1.7

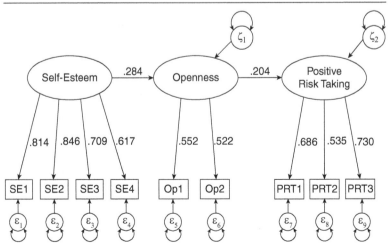

pedagogical, accessible, and comprehensive and includes examples, exercises, and recent advances that make this textbook excellent for SEM users of all levels. In addition, for more sophisticated SEM researchers, Bollen's (1989) classical text on SEM is highly recommended for researchers interested in the advanced SEM for its complete coverage of the matrix algebra underlying SEM.

Many SEM textbooks are written specifically for use with a companion software package, such as Acock (2013) for Stata, Kelloway (2014) for Mplus, Beaujean (2014) for R, and Byrne (2016) for AMOS. These texts effectively acquaint the researcher to both SEM and the particular software package. However, care must be taken with texts tied to a particular software package, as software is updated at exceptional rates, as are the various SEM techniques supported by a particular software program.

Meta-Analysis of Correlation Matrices

Summary

Meta-analysis is a statistical procedure in which the researcher pools effects from multiple independent studies. By synthesizing multiple studies into a meta-analysis, researchers can achieve the following goals: both quantitatively and qualitatively summarize a body of literature, estimate average effect sizes more accurately than a single study is able to achieve, explore publication bias in the literature of an effect, and investigate why reported effects differ among studies (so-called heterogeneity of effects)— for example, due to between-study differences in sample demographics, duration of study, or types of assessments used. Meta-analytic techniques have been developed for pooling means, odds ratios, regression coefficients, and correlation coefficients. More recently, meta-analytic techniques for pooling correlation matrices have been developed with two goals: first, to accurately estimate a pooled correlation matrix for the variables of interest (or perhaps several matrices, if particular heterogeneity of effects warrants estimating separate correlation matrices across groups), and second, to conduct SEM on the pooled correlation matrix or matrices using meta-analytic SEM (MASEM; e.g., Cheung, 2014; Cheung & Chan, 2005).

Earlier work on estimating pooled correlation matrices used meta-analytic techniques for individual, bivariate correlations. Common methods were to average either the raw correlation or the Fisher-transformed z value across independent studies for each bivariate relationship. Each correlation in the matrix is pooled in this way, and each correlation potentially may be based on a different number of samples. These bivariate methods—applied to a multivariate correlation matrix—are simple to implement and adequate for computing a pooled correlation matrix. However, for several reasons, they are not recommended if the researcher plans to implement MASEM. First, if some studies report only some, not all, of the variables of interest, then each correlation may have been computed using different samples (likely of different sizes), so there is no guarantee the pooled matrix is PSD, and there is no single, obvious sample size to input to estimate accurate standard errors in SEM. Second, even if the pooled matrix is based on studies each of which includes all variables, conducting SEM on the averaged correlations neglects the heterogeneity in observed correlations across the included studies, and that heterogeneity is likely to be of theoretical interest. Third, pooling each correlation separately ignores the natural dependence among multiple correlations reported in a single study.

As MASEM has gained popularity, better methods for pooling correlation matrices have been developed that account for the dependence among correlations within studies. In particular, MASEM methods that use generalized least squares or multiple-group SEM (Cheung & Chan, 2005) to combine correlation matrices can also estimate and accommodate heterogeneity in correlations across studies in the final SEM. These approaches also improve on using bivariate methods by reducing confusion about appropriate sample size for SEM.

Recent MASEM Examples

We refer researchers interested in conducting MASEM to two recent empirical examples that demonstrate both the meta-analytic process (including steps in a literature search and the synthesis process) and the MASEM-fitting process (including details of the model and interpreting estimates and model fit). Quinn and Wagner (2018) used MASEM, fit to a pooled correlation matrix, to explore the factor structure of reading comprehension. This example may be of particular interest for researchers who wish to use MASEM to help with scale development. Protogerou et al. (2018) used MASEM, also fit to a pooled correlation matrix, to investigate mediators of past and present condom use among sub-Saharan African young adults and provided an excellent example for researchers interested in using MASEM to look at relationships among several study variables.

Resources for Software and Additional Readings

As MASEM is a relatively new technique, there are few software options to conduct MASEM with ease. Older applications of MASEM first pooled correlation matrices in one program, using any of the previously discussed methods, before conducting SEM on the pooled correlation matrix in any SEM software. As indicated, this approach has issues. At present, the

60

metaSEM package in R is the most up-to-date option for substantive researchers to pool correlation matrices and conduct MASEM in one program. Various options for pooling the matrices and for fitting the SEM models are provided. For details about the software, including a demonstration, see Cheung (2014). The main drawback of metaSEM is the lack of standard meta-analytic output that typically accompanies meta-analyses. For instance, this package does not provide funnel plots, which can be used to estimate the presence of publication bias. These plots may be of interest for a single or several correlations estimated. Forest plots are also not supported in the metaSEM package; these plots show the distribution of estimates for a single correlation across studies. These types of meta-analytic plots and output are readily available in Stata; RevMan (a free, dedicated meta-analytic program); and R (using the metafor package). We note that software to support meta-analysis of correlation matrices is early in its development and may, quite possibly, be further advanced by the time readers of this book wish to access such software. The software listed in this paragraph is, obviously, a starting point, but may not be state of the art beyond the immediate future.

Summary

Bivariate correlation formulas were first developed in the late 1800s by Pearson and Galton. The first factor analysis approach, used to study the correlation matrix among intelligence subscales, was developed only shortly afterward by Spearman in 1904 (see Figure 1.2). Path analysis was first proposed by Wright in 1934. Obviously, the idea of defining models of correlation matrices by decomposing those matrices into informative factors and components has been in development for almost as long as the computation of correlation coefficients themselves.

SEM was developed in the second half of the 20th century by Jöreskog and others (e.g., Jöreskog, 1970), and it has developed into one of the most powerful analytic tools in the social sciences. The SEM approach, fit to a correlation or covariance matrix, blends factor analysis as a measurement model and regression analysis as a structural (predictive) model and provides many statistical advantages. First, SEM provides a methodology that can simultaneously estimate many different statistical features (instead of, say, first estimating a factor analysis model based on a correlation matrix, followed by several different regression models that attempt to somehow combine the results of the factor analysis and the regressions); the advantages to simultaneous estimation are not only efficiency but also control of Type I error rates. Second, standard regression models in the general linear model tradition include a component for measurement error in the dependent variable, but the independent variables are assumed to be measured without error. SEM allows the researcher to define error components for all variables and further allows the errors to be correlated (or not) across variables. Third, the ability to evaluate the fit of an SEM to the original correlation/covariance matrix across the whole model, or within subcomponents of the model, and with a number of different types of fit statistics, provides a great deal of flexibility for the applied researcher using SEM.

Finally, using SEM as a model of correlation matrices provides entry into many other modern statistical methods. SEM has been linked mathematically to multilevel modeling, to growth curve modeling, to meta-analysis (as we illustrate in the last section), and to many other modeling methods. SEM software can be used to estimate models of correlation matrices within any of these contexts, and many others as well.

Chapter 5

GRAPHING CORRELATION MATRICES

In this chapter, we describe and illustrate a number of graphical methods that have been developed to visually portray the information contained in a correlation matrix. The use of graphical methods has increased systematically since Tukey (1977) defined graphical analysis as one of the four hallmarks of exploratory data analysis (along with robustness and resistance, transformations, and residual analysis). Tukey (1977, pp. 1–3) drew a strong analogy between a research project and a courtroom setting, both of which are devoted to using empirical evidence to understand processes that are unclear or deceiving. He drew an even tighter analogy between the kind of evidence that is collected during a crime investigation—fingerprints, shoeprints, video evidence, document evaluation, and so on—and the kind of data- and model-based evidence that researchers use to develop hypotheses and models about how the world works—residuals, transformations, and, especially, graphical methods. In other words, many goals and methods are shared by the two apparently disparate professions—the detective using standard evidence to attempt to solve a crime and the researcher using graphs, residuals, and other exploratory approaches to identify, clarify, elaborate, and illustrate a research finding.

We use several quotes from Tukey about graphing to help motivate this chapter on graphing correlation matrices:

Graphs are friendly. (Tukey, 1977, p. 157)

The greatest value of a picture is when it forces us to notice what we never expected to see. (Tukey, 1977, p. vi)

As yet I know of no person or group that is taking nearly adequate advantage of the graphical potentialities of the computer. (Tukey, 1965, p. 26)

The point for the reader to take from these quotes is that complex data can be effectively summarized by using graphs. And complex data can usually be better understood by studying one or more graphs than through any medium that does not include visualization. One of the most useful, and also complex, statistical structures is the correlation matrix, which can be graphed in a number of different ways. As Tukey (1977) notes, "Different graphs show us quite different aspects of the same data. . . . There is no more reason to expect one graph to 'tell all' than to expect one number to do the same" (p. 157).

There are several reasons that graphs of correlation matrices are challenging, and then ultimately very valuable. The first difficulty is that correlation matrices are defined (potentially) in many dimensions. If there are 10 variables and 100 subjects, then the technical statistician will recognize either 100 subjects defined as points in a 10-dimensional variable space, or 10 variable vectors defined in a 100-dimensional subject space. But our visual information system is limited to three dimensions (or, perhaps, four dimensions, if we count a three-dimensional pattern changing over time, although the time dimension functions differently than the three space dimensions). However, it is important for both the expert and the novice to recognize that 10-dimensional and 100-dimensional spaces exist, conceptually and mathematically, even if we cannot literally "see them." Creative ways have been developed in which we can visualize correlation matrices graphically, even those defined in higher than three/four dimensions.

The second challenge to producing graphs of correlation matrices is the production side of the first challenge. Just as we are limited in our ability to see beyond three dimensions, we are also limited in our ability to render high dimensions in a way that can be visually inspected. It is straightforward to render three dimensions on a two-dimensional surface; artists have been doing that for much longer than graph making has existed. But showing high-dimensional data spaces—within which are bivariate correlations that can be combined into a correlation matrix—is a difficult production process. Fortunately, creative approaches to showing high-dimensional patterns have been developed, as we will present and illustrate in this chapter.

The third challenge, related to the first, is addressing the question—the choice—of whether subject points are defined in a variable space, or variable vectors are defined in a subject space. Mathematically, the two conceptualizations are equivalent (mathematicians would say they are "isomorphic"). But in terms of how data are presented graphically, each has advantages and disadvantages (some of which we will identify in this chapter and in the next chapter as well).

The fourth challenge in graphing a correlation matrix reflects one of the primary themes of this book: A correlation matrix is much more than just an arrangement of bivariate correlations. A graph of a correlation matrix should ideally reflect the "much more" built into that statement, the complex multivariate intercorrelations that result in constraints defined across the multiple bivariate correlations. But many of the methods to graphically portray a correlation matrix are inherently bivariate and do not logically or technically address those constraints or the higher-level multivariate structure across multiple variables. Although multiple bivariate graphs can be very helpful to a data analyst, they stop short of actually capturing the deeper structure of the correlation matrix. We will discuss this challenge at several points as we present specific techniques, and then we will return to and further discuss this challenge in our conclusion.

The fifth and final challenge involves software, and this challenge is actually a "challenge of riches." Tukey's 1965 quote above about "the graphical potentialities of the computer" has been realized; the statement itself is now quite dated, and largely no longer applicable, which undoubtedly would delight the original author. Many software approaches exist to manage and then ultimately graph data, including correlation matrices. Major software systems such as SAS and SPSS specialize in graphical methods (SAS has both a separate module called SAS/GRAPH and a standalone separate package called JMP, both of which are dedicated to production-level and publication-quality graphs for research, business, education, etc.). The freeware system, R, has outstanding graphical capabilities, as we illustrate in this chapter. Many other dedicated graphics packages exist, some of which are designed to produce specialty graphs related to a particular need. The ggplot package (now actually ggplot2, both of which are R packages) is an exemplary product within this category that was developed from the principles defined in Wilkinson's (1999) *Grammar of Graphics*. The lattice package is another high-level graphical software system in R; this package is a later-generation version of the Trellis graphics system that was originally developed for S and S-Plus.

We organize this chapter approximately in relation to the sophistication of graphical applications that can produce graphs of correlation matrices. We begin with some treatment of how to graph individual correlations themselves, which will set the stage for graphing correlation matrices. Following, we present some of the most basic (and most useful) correlation matrix graphs, with particular focus on one of the simplest, the scatterplot matrix. Then, we present more complex and specialized graphical methods for correlation matrices. In each case, we illustrate the graphical application in relation to data examples used in this book.

Graphing Correlations

Although correlation matrices are much more than simply a collection of correlation coefficients (see Chapter 2), at least some of the graphing approaches to correlations can be easily expanded into correlation matrices simply by applying them to pairs of variables. Thus, in this section, we discuss how correlations are typically graphed. Then, in later sections, we extend that treatment to correlation matrices. Some graphical methods graph correlation matrices by defining pairwise graphs of correlations.

66

Figure 5.1 A Scatterplot of Two Variables, X_1 and X_2

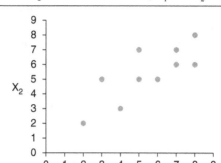

Only a very few other approaches are available that graph the deeper multivariate features of the correlation matrix.

The basic method used to plot bivariate data in variable space (i.e., with orthogonal—perpendicular—axes that represent the two variables) is called the scatterplot, which may, in fact be, the most popular and widely used graphical method for portraying data that exists (see Jacoby, 1997, for expansive treatment of the scatterplot). In Figure 5.1, we present 10 subjects' scores on two variables and plot those subjects' ordered pairs in a scatterplot.

Several features of this scatterplot are relevant and important. First, the two variables, X_1 and X_2, define the orthogonal axes of the two-dimensional space. Second, each subject is represented as an ordered pair, defined as a point, within the variable space. Third, these two particular variables have means and standard deviations that are not identical, but in this particular case, they are relatively similar; that is, the two variables were measured on approximately the same scale. However, this is not generally the case; consider plotting SAT (Scholastic Aptitude Test) against high school GPA (grade point average), or height in inches against weight in pounds. Scatterplots can be plotted in the raw units of the variables, or as standardized transformations of the two variables, in which each variable is adjusted using the z-score formula to have a mean of 0 and a standard deviation of 1 (and hence, the scales of the two variables are equated).

Note that the paragraph above does not use the word "correlation." A bivariate scatterplot, however, is a powerful visualization tool for representing correlations. Many scatterplots (including the one above) force the eye to follow a pattern; in the example in Figure 5.1, the pattern of the data moves from lower left to upper right—that is, an increasing pattern.

This type of pattern implies a positive correlation in the data; that is, as scores on the first variable increase, scores on the second variable tend to increase, and as the first variable scores decrease, the second variable scores tend to decrease. A pattern of data movement from upper left to lower right implies a negative correlation in the data. A pattern of data movement that is relatively flat implies a correlation coefficient of around 0.

This discussion implies that there is a line (or multiple lines) that could be passed through the data as summaries of the bivariate data pattern. The most popular way to define such a line is by using least squares regression (e.g., Lewis-Beck & Lewis-Beck, 1980) as a criterion to define a line through a pattern of data points in a two-dimensional scatterplot (there are other criteria that can be used, which usually will result in slightly different lines). The tightness with which the points cluster about such a line is indicative of the size of the correlation (though solely relying on clustering to interpret the size of a correlation can be deceptive; see Loh, 1987). A horizontal regression line with little clustering reflects a correlation around 0, whereas a regression line sloped upward toward the right (as in Figure 5.1) with the points fairly tightly clustered around the line reflects a correlation higher than 0 (in the example, $r = .79$). If the points are so patterned that they fall exactly on a line, then $r = \pm 1$, the largest possible correlation reflecting a perfect linear relationship between the two variables.

But there are risks in using a scatterplot to visualize the correlation between two variables. The first risk is that how the scales of measurement are portrayed in the graph can distort correlations and can cause the size of the correlation to be apparently magnified or suppressed. If this is a concern, plotting the standardized variables is an excellent solution. A second risk occurs in relation to nonlinear patterns in the bivariate data structure. For example, the data in Figure 5.2 have a perfect relation underlying this pattern—$Y = X^2$—but $r_{YX} = 0$. Anscombe (1973) famously developed four completely different patterns of relationships that were portrayed in a scatterplot as defining different types of bivariate relationships, but which had exactly the same underlying correlation. The take-home message from both of these types of risks is that different information emerges from studying the pattern in a scatterplot and from looking at the value of a correlation. Shrewd data analysts will look both at a scatterplot and at the value of the correlation in an attempt to understand the patterns built into a bivariate data structure.

Another way to visualize a single correlation between two variables is to turn the scatterplot inside out (see Rodgers & Nicewander, 1988, for elaboration). In other words, if we let the subjects define the orthogonal axes (e.g., a 10-dimensional space, using the data in Figure 5.1), we can plot the variables as vectors (directed line segments) within that space; each vector

68

Figure 5.2 A Quadratic Relationship Between Y and X, With an Equation That Perfectly Predicts Y From X ($Y = X^2$), But $r_{YX} = 0$

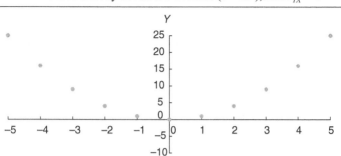

starts at the origin and finishes at the value of the ordered "10-tuple" defining the variable. Thus, for X_1 in Figure 5.1, the vector starts at (0, 0, 0, 0, 0, 0, 0, 0, 0, 0) and finishes at (3, 2, 8, 5, 7, 5, 6, 8, 7, 4). Though we cannot "see" in 10-dimensional space, the two vectors emanating from a common origin define a two-dimensional plane (or, possibly, a one-dimensional line if they are colinear) in which we can see the variables (this plane is a subspace within the 10-dimensional space, one that can be visualized).

If the variables are standardized, then at least two important adjustments occur. First, the origin of the space is shifted to the point representing the means of all the variables. Second, the angle between the variable vectors is related to the correlation; specifically, r is the cosine of the angle between the two variable vectors. (The cosine function is the standard trigonometric function that most students studied in high school; virtually all calculators have a button for the cosine—and sine, and tangent, and other—functions.) Note that two vectors lying at right angles have a cosine (90°) = 0.0, and $r = 0.0$. Two variables that fall on the same line and point in the same direction have cosine (0°) = 1.0, and $r = 1.0$. Two variables that fall on the same line, but point in opposite directions, have cosine (180°) = −1.0, and $r = -1.0$. Two variables with vectors with a 45° angle between them have cosine (45°) = .707, and $r = .707$. In Figure 5.3, we present examples of variable vectors corresponding to this second interpretation. The X_1 and X_2 variables in Figure 5.1 have $r_{12} = .79$; the cosine that produces such a correlation has an angle of 37.8°, which is illustrated in Figure 5.3.

This second perspective, visualizing a correlation in relation to the angle between the variable vectors, may at first appear mathematically complex. Once the reader drops the concern over the (hidden) high-dimensional space that is in the background, there are several advantages of this second perspective compared with the standard scatterplot. First, the scale of

Figure 5.3 Three Examples of Pairs of Variable Vectors Lying in Subject Space, in Which the Cosine of the Angle Between Them Is the Correlation

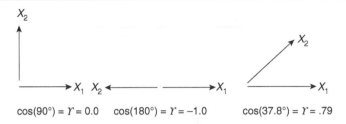

measurement issue discussed above, which can distort correlations in scatterplots, is irrelevant when we standardize both the variables (which equates their scale of measurement). Second, there is a natural link between the correlation and the geometry; uncorrelated variable vectors are perpendicular, perfectly correlated variables lie on the same line, and so on, just as we conceptually feel that they should. Third, this perspective gives a natural way to extend the relationship built into a single correlation into the many correlations that lie in a correlation matrix. Each variable underlying the correlation matrix is simply a variable vector in subject space, emanating from the same origin. If we have five variables, and thus a 5×5 correlation matrix with 10 separate correlations—r_{12}, r_{13},..., r_{23},..., r_{45}—these can be conceptualized as the 10 angles between pairs of five variable vectors, with each correlation defined as the particular cosine of the angle between the vectors. Once the reader grasps the elegance of this visualization, they will appreciate that a data space is simply a pattern of vectors that project out into the space, like those in Figure 5.3, except with more vectors and more dimensions. Then, the correlation matrix has a very natural link to this data space, through the size of the angles between the variable vectors. It is of historical interest, and also of pedagogical value, to note that this high-dimensional subject space allowed Sir Ronald Fisher in the early 20th century to conceptualize statistical patterns that had not previously been recognized (see Box, 1978, for discussion of how Fisher used this space), and ultimately led to the development of statistical tests for the correlation coefficient, as well as the analysis of variance.

Graphing Correlation Matrices

In the last paragraph of the previous section, we began the transition from graphing correlations to graphing correlation matrices. In this section, we will present a number of different approaches that have been developed to graph a correlation matrix. Several are natural applications of methods described above to graph single correlations, except extended to patterns of many correlations. Others are uniquely defined in relation to correlation matrices per se. We begin with what may be the simplest, yet also perhaps the most powerful, data visualization method ever developed for visualizing patterns in correlation matrices: the scatterplot matrix.

The Scatterplot Matrix

Cleveland (1993), who presented a large number of useful techniques for data visualization, was especially enamored with the scatterplot matrix: "An award should be given for the invention of the *scatterplot matrix*, but the inventor is unknown" (p. 274). He noted that Chambers et al. (1983) had the earliest written description of the scatterplot matrix, but the historical trail is lost earlier than that. The implication from Cleveland is that this simple yet powerful visualization tool may have simply evolved as a natural and recognized need.

The scatterplot matrix is very simple to define, and its relationship to the correlation matrix is obvious from its definition. A scatterplot matrix is a pairwise plot of all the scatterplots that underlie a correlation matrix. The scatterplot matrix is to a correlation matrix exactly what a scatterplot is to a single correlation. Furthermore, the scatterplot matrix is organized just as the correlation matrix is organized, as a row by column matrix of scatterplots; a given scatterplot is the one that underlies the correlation that is in the same location in the correlation matrix. Like the correlation, corresponding off-diagonal elements of a scatterplot matrix are redundant; each plot is simply the corresponding plot, reflected from the upper (lower) to the lower (upper) triangle of the space. Thus, some software systems provide only the upper (or lower) triangle of a scatterplot matrix. We illustrate in Figure 5.4, which shows the scatterplot matrix of the state-level data that are underlying the correlation matrix originally presented in Table 1.4. Each graph, produced using the *pair()* function in R, shows the scatterplot for each pair of the five variables. We will provide some interpretations for these patterns in the next section, after we add an important enhancement that helps with understanding these patterns.

The Scatterplot Matrix, Enhanced

Many ways exist to annotate or enhance a scatterplot matrix in order to create an even more informative portrayal. A number of those are treated and demonstrated in Cleveland's (1993) book *Visualizing Data*. We will review the two most important of those in this section.

71

Figure 5.4 A Scatterplot Matrix of the Five Variables From the U.S. State Data From Chapter 1, Table 1.4

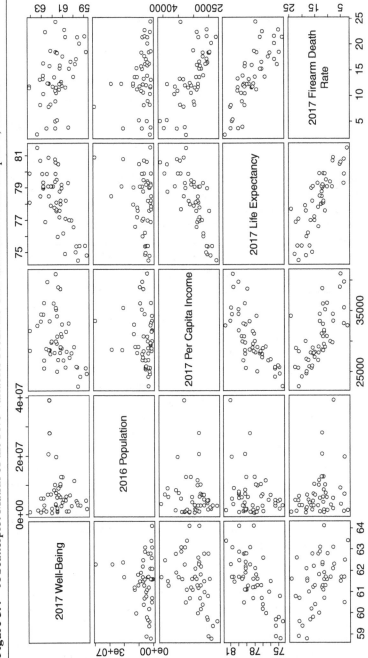

The first approach to enhance the interpretation of a scatterplot matrix (or single scatterplots for that matter) is a superimposed linear model of the data. There are a number of options, most of which provide relatively similar lines that pass through the data. The most common linear model in applied research settings is a regression line—a straight line based on the least squares criterion. Regression (and other linear) models work well to enhance scatterplot matrices when the scatterplots have an underlying approximate linear structure.

Alternatively, a *loess* line can be superimposed on each scatterplot in the scatterplot matrix. The portmanteau *loess* is short for "local regression" (Cleveland, 1993, p. 94; also see Jacoby, 1997); the method is also often called LOWESS, an acronym for "LOcally WEighted Scatterplot Smoothing." The *loess* method was originally developed by Cleveland (1979), who presented it as a sequential fit of local functions (often regression lines, and the user can choose higher-order polynomial regressions instead of linear regression) in a left-to-right direction to small portions of the data set. For example, in a plot with 100 data points, for a lambda parameter of 1 (indicating linear regression) and an alpha parameter of .05 (reflecting that 5% of the data are used in any given regression), the first five points from left to right have a regression line fit to them, and the third point is replaced with the prediction obtained from that line. Then the second through sixth points are fit, and the fourth point is replaced with the prediction for the middle point. After running the procedure for every set of five points, moving left to right one at a time, the predictions are connected into a *loess* line, and a nonlinear pattern that smooths the data structure is obtained. An example of *loess* lines is shown in Figure 5.5, which is the same scatterplot matrix as in Figure 5.4, but with the *loess* curves added as enhancements by superimposing the curve on the scatterplot (and note that we only include the upper triangle).

The *loess* approach is perhaps the most popular, but it is only one among many different smoothing approaches. Alternatives include spline smoothing, polynomial smoothing, and median/mean smoothing. In each case, a defined fraction of the data is fit with a model, and the model value points are then connected into a nonlinear structure that fits the data—at the local as well as global levels.

Adding lines or curves substantially improves our ability to interpret the patterns that underlie the scatterplots. For example, in Figure 5.5, we can see several interesting patterns. First, well-being is relatively unrelated to population, but it is strongly associated in an approximately linear manner with life expectancy. The relation of well-being to per capita income is more subtle; well-being increases with income up to a certain point, and then levels off. Income and life expectancy have a strong positive relationship.

73

Figure 5.5 A Scatterplot Matrix of the Upper Triangle From Figure 5.4, With Scatterplot Smoothing (*loess*, or LOWESS) Enhancement to Facilitate Interpretation

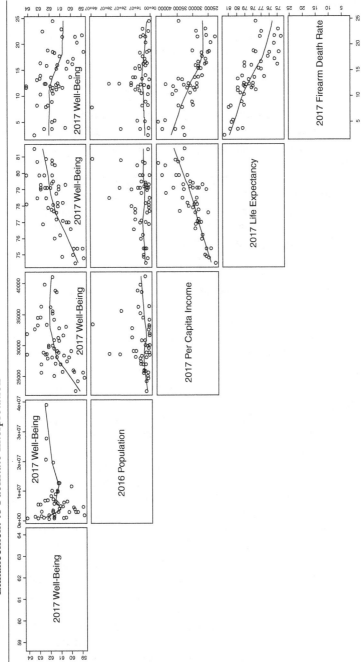

Finally, life expectancy and death from firearms have a strong negative association. We note that, of course, none of these relationships can be construed as necessarily causal, as correlation is a necessary but not a sufficient condition for inferring causality between two variables.

Corrgrams Using the corrplot Package in R

The corrplot package in R appears to have emerged from Friendly (2002), who defined the term *corrgram* as a set of techniques that (a) render the sign and magnitude of a correlation in a correlation matrix and (b) potentially reorder the variables to place similar variables near one another. He defined five methods to graphically portray a correlation: numbers, circles, ellipses, bars, and shading. Corrgrams also are called correlograms on occasion.

The corrplot package expands on the options originally defined by Friendly (2002). Using the basic orientation of a scatterplot matrix, corrgrams produced with corrplot show the pattern underlying each pair of variables using one of seven geometric structures: circles, squares, ellipses, numbers, shade, colors, and pie charts. The upper diagonal, lower diagonal, or the whole matrix can be displayed graphically, and the options may be separately defined in the upper and/or lower triangles.

The reader may wish to know how the geometric shapes—circles, squares, ellipses, and so on—are used to represent correlation coefficients within a correlation matrix by the corrplot package. Most of the mappings from the correlation coefficient into the geometric shape are highly intuitive and are described in detail in several articles in the statistical literature (in particular, the corrplot documentation cites Friendly, 2002, and Murdoch & Chow, 1996). For example, both circles and squares are represented so that the area of the circle/square is proportional to the absolute value of the correlation coefficient. Ellipses have the advantage that they can be oriented as tilted upward or downward to reflect the sign of the correlation. Furthermore, the ellipse is "shaped to be contours of a bivariate normal distribution with unit variances and correlation ρ, with the contour tangent to a unit square" (Murdoch & Chow, 1996, p. 178). The exact formula used to calculate a given ellipse can be found in Murdoch and Chow (1996) below this quote. Conceptually, as the absolute value of the correlation increases then the ellipse is tilted upward and is "slimmer" for positive correlation, and downward and is "slimmer" for negative correlation. At the limits, the ellipse becomes a straight line oriented at 45° and −45° for $r = 1.0$ and $r = -1.0$, respectively, and the ellipse is a perfect circle for $r = 0$. The other options—numbers, shade, colors, and pie charts—are even more intuitive (and, in some cases, require a legend within the figure). These options are illustrated using examples below.

As is the case with many graphical methods, corrgrams are easier to show by example than to describe. In Figure 5.6, we plot corrgrams from the Elgar (2010) country health data from Table 1.3. This correlation matrix is highly disparate, ranging from large negative to small negative to small positive to large positive correlations. The first graph in Figure 5.6 shows the ellipse option. Murdoch and Chow (1996) were interested in graphical portrayals of very large correlation matrices and recommended the use of ellipses in such settings (with, e.g., greater than 40 variables; they noted that scatterplot matrices are most effective for between 10 and 20 variables). Although the correlation matrix in Figure 5.6 is relatively small, these conventions can be observed in the first corrgram in Figure 5.6.

The second corrgram in Figure 5.6 shows the circle option in the upper triangle, combined with the actual correlation values in the lower triangle (note that corrplot produces more than just the value of the correlations; by default, the numeric values are also shaded to help the viewer recognize high and low correlations). The correlation matrix symmetry can be easily viewed, as there is obvious redundancy between upper triangle circles and lower triangle correlation values of the corrgram. Presenting both a geometric interpretation and the value of the correlation allows careful inspection of these two different ways to present correlation values.

Finally, in the third corrgram of Figure 5.6, we present another example using the Elgar (2010) health data from Table 1.3. The correlations are represented with the square option in the upper triangle, and the pie chart option in the lower triangle. Using the square option for the upper triangle—which uses color to distinguish the size of correlations—results in corrgrams that resemble a heat map, which will be treated separately in the next section. Heat maps can be presented more clearly using the color option in corrplot.

Heat Maps

Heat maps use color to represent data values, and these are used in many data visualization applications besides correlation matrices. Using the color option in the corrplot package produces a heat map, as shown (using greyscale) in the top triangle of the third corrgram in Figure 5.6. In the case of a correlation matrix, a heat map portrays the size of a correlation as a particular color (or shading, in grayscale graphs). As a result, a researcher studying a correlation matrix using a heat map can use visual inspection compared with the type of cognitive inspection required in studying the actual numerical values in a correlation matrix (or even the geometric figures in the other options in the corrplot package).

76

Figure 5.6   These Three Corrgrams Present the Correlation Matrix
Showing Country Health Patterns for 33 Countries (Elgar,
2010) From Table 1.3

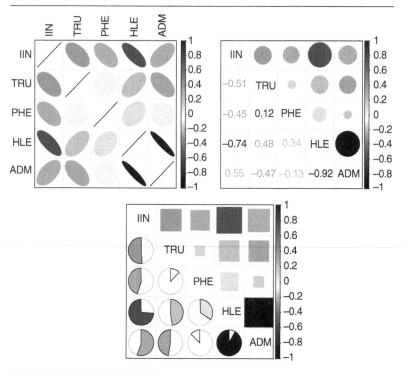

Note: The first corrgram presents the ellipse option, the second shows the circle option in
the upper triangle and the correlations themselves in the lower triangle, and the third presents the square option in the upper triangle and the pie chart option in the lower triangle.
The second and third corrgrams demonstrate how separate options can be used in the upper
and lower triangles. IIN = income inequality; TRU = trust; PHE = public health expenditures; HLE = healthy life expectancy; ADM = adult mortality.

A number of software systems specialize in producing heat maps. SAS/
GRAPH is a high-end graphical system that does so, and Excel is a lower-
end (but easier to use) system that also produces heat maps. R has effective heat map capabilities through corrplot and ggplot2. The business
graphical software Tableau also produces high-quality heat maps.
Two different heat maps of the country health data from Table 1.3 are
shown in Figure 5.7 (note that the left-hand graph reproduces one of the
graphs from Figure 5.6).

Figure 5.7 Two Different Approaches to Portraying a Heat Map (Using the corrplot Package in R) of the Country Health Data (Elgar, 2010) From Table 1.3

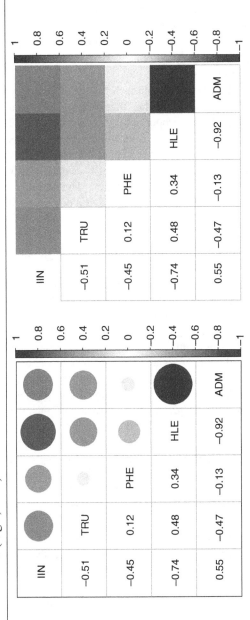

Note: IIN = income inequality; TRU = trust; PHE = public health expenditures; HLE = healthy life expectancy; ADM = adult mortality.

Parallel Coordinate Plots

Coordinate axes are routinely defined as perpendicular (orthogonal) to one another, so much so that it is easy to assume that this is the "correct" (or even the only) way to define coordinate axes. Yet there exist other axis representations. In other words, using perpendicular axes is a decision (which is often implicit).

A different and quite useful approach that can effectively portray a correlation matrix uses parallel, rather than perpendicular, coordinate axes. The method originally was proposed by Inselberg (1985) and further developed by Wegman (1990). In a parallel coordinate plot, all axes are parallel to one another (and can be portrayed either horizontally or vertically). The potential for different scales must necessarily be handled, for example, by equating the scales through standardization. The parallel coordinate plot defined by two parallel lines (e.g., X and Y) is a counterpart to a scatterplot defined in relation to orthogonal axes; that is, it shows the relationship between two variables.

For a given subject, the score on X and the score on Y are connected by a line segment from the top (say, X) to the bottom (say, Y) horizontal axis— or, equivalently, from the left (say, X) to the right (say, Y) axis. Across many observations (respondents, individuals, countries, hospitals, etc.), the general correlational pattern can be observed. If the connections between X and Y are usually nearly vertical for horizontal axes, or nearly horizontal for vertical axes, that would imply that observations define approximately the same standardized value of X and Y and a high positive correlation between X and Y. If the connections are from low values on X to high values on Y, and vice versa (characterized by highly angled line segments), this pattern would be reflective of a high negative correlation. Thus, two parallel lines can be used to represent a single correlation between two variables. By using multiple parallel lines, multiple variables can be represented; that is, multiple lines can be used to (at least partially) represent a correlation matrix. As in many graphical procedures, the verbal description just presented is far weaker a demonstration than the presentation of actual data using a graph. We present several such examples, using the *parcoord*() function in the MASS package in R.

First, we present a very simple parallel coordinate plot in Figure 5.8 using the 10 data points presented in Figure 5.1 at the beginning of this chapter. In this plot, X_1 is the left vertical axis and X_2 is the right vertical axis. The scale on each axis is defined from 2 (the minimum for each variable) to 8 (the maximum for each variable). The ordered pairs that were represented in the scatterplot in Figure 5.1 are now lines, stretching from the X_1 vertical axis to the X_2 vertical axis. The relatively high positive correlation among X_1 and X_2 ($r_{12} = .79$) is reflected in the parallel coordinate

79

Figure 5.8 A Parallel Coordinate Plot of the Raw Data ($N = 10$) Shown in Figure 5.1

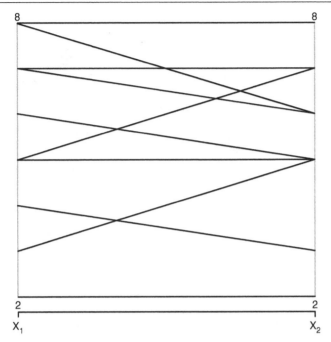

plot in that the lines generally stay in the same part of the graph; the lines that start high on X_1 stay fairly high on X_2, and vice versa.

In Figure 5.9, we present a much more sophisticated version of a parallel coordinate plot. Parallel coordinate plots (like scatterplot matrices) plot raw data (the patterns that underlie a correlation matrix), and so we use the U.S. state data (which we presented as a scatterplot matrix in Figure 5.4). At the top of Figure 5.9 is a parallel coordinate plot for all five of the U.S. state data variables. We focus on two pairs of variables to further illustrate the results of a parallel coordinate plot. In the scatterplot matrix, we noted the strong positive correlation among income and life expectancy. In the middle plot, this correlation is graphed, and the strong positivity of the correlation is illustrated by most of the data lines being approximately horizontal. On the other hand, we noted a strong negative correlation between life expectancy and deaths from guns. In the bottom plot, this correlation is graphed, and the strong negative correlation is illustrated by consistent crossed lines. That is, high values on one variable correspond to low values on the other, and vice versa.

80

Figure 5.9 Parallel Coordinate Plots for the U.S. State Data from Table 1.4

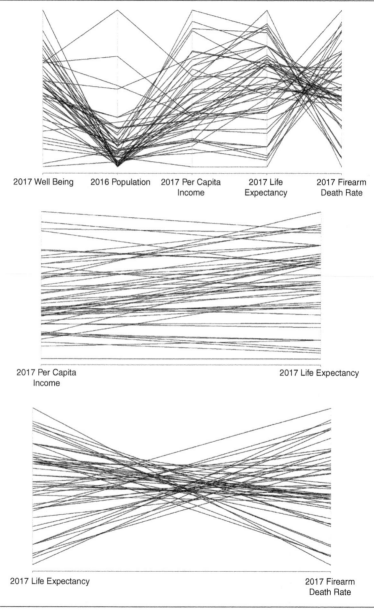

Note: The first graph shows all five variables, the second graph shows pairs of variables illustrating positive correlation, and the third graph shows pairs of variables illustrating negative correlation.

There is one obvious disadvantage to parallel coordinate plots: It is difficult to display correlations between all pairs of the variables in a correlation matrix in a single plot. Because only the contiguous parallel lines can be easily used to present a correlation, the order of the lines (i.e., the order of the variables) matters. Different orderings would show different patterns within the correlation matrix. Dawson et al. (1997) provided one conceptual solution to this problem by applying parallel coordinate plots to repeated measures data, which have a natural ordering in relation to time. Alternatively, several methods have been developed in the past to define an empirical ordering of the variables; these will be briefly reviewed in the next section.

Eigenvector Plots

Friendly (2002) presented eigenvector plots as graphical portrayals of correlation matrices. Apparently, little application of this method has been developed, and few examples can be found in the literature. However, we want to emphasize the value of this kind of graphical portrayal. Eigenvalues and eigenvectors account for the multivariate structure of the correlation matrix. The previous graphical methods presented in this chapter—including the most popular methods such as scatterplot matrices and corrgrams—involve the multiple application of bivariate graphical methods. These approaches treat correlation matrices as though they are a collection of correlations. Although the graphical inspection of such bivariate graphs can provide valuable insights for researchers, we have repeatedly emphasized that correlation matrices contain information that is deeper and more nuanced than that represented in the multiple bivariate relationships. The eigenvalue plot of a correlation matrix begins to capture that nuance. However, it is limited as well in ways that we will review in the conclusion to this section.

As we discussed in Chapter 2, the eigenvectors and eigenvalues are critical features of a correlation matrix. Many matrices filled with correlations are not true correlation matrices. (In fact, a very small proportion of such matrices beyond the 3×3 case are true correlation matrices; most are pseudo-correlation matrices. See Chapter 6.) To remind the reader of how a pseudo-correlation matrix can be identified, remember that in true correlation matrices all the eigenvalues are nonnegative. If one or more of the eigenvalues are negative, then there exist no underlying data values that could have produced the apparent correlation matrix.

As an example, in the correlation matrix in Table 1.6, representing fertility planning among 7,000 NLSY respondents, this five-variable correlation matrix has eigenvalues of 2.2, 1.6, 0.7, 0.5, and 0.1. (Note that this matrix was slightly adjusted in Table 1.6 to demonstrate the difference between true and pseudo-correlation matrices, and as described in Chapter 2; the version of the correlation matrix that is used in this chapter is the original unadjusted correlation matrix, with the $r_{12} = .876$ entries adjusted back to their original values of .756.) Because the eigenvalues are all positive, this is a true correlation matrix (and the determinant is positive, as it must be for this to be a true correlation matrix, $|\boldsymbol{R}| = 0.10$). Furthermore, there are two relatively large eigenvalues. We can compute the percentage of variance associated with the direction of the first eigenvector by taking the first eigenvalue divided by the sum of all the eigenvalues, $2.2/(2.2 + 1.6 + 0.7 + 0.5 + 0.1) = 43\%$. The second eigenvector's direction is associated with $1.6/(2.2 + 1.6 + 0.7 + 0.5 + 0.1) = 31\%$ of the overall variance. Thus, the two orthogonal directions identified by the first two eigenvectors account for 74% of the overall variance across all five of the variables.

Friendly (2002) recommended plotting each variable as a point represented by the dominant eigenvectors. For example, the five eigenvectors from the correlation matrix of Table 1.6 are presented in Table 5.1; the first two eigenvectors, which we will use as the dominant eigenvectors, are bolded in the table. In Figure 5.10, we present an eigenvector plot of the first two eigenvectors and the five variables that are represented in the correlation matrix in Table 5.1 (this is a plot of the first two columns of Table 5.1, with a point defined by the ordered pair in each row). This plot was produced in SAS, using PROC SGPLOT, which is an SAS plotting procedure designed to plot and label vector representations. It is clear from this plot that the fertility attitude variables Expect79, Ideal79, and Ideal82 are relatively similar to one another and are important in defining the dominant direction defined by the first eigenvector (in many treatments, this direction is referred to as the first principal component). The fertility behavior variables—Number80 and Number04—are relatively similar to one another and are important in defining the second direction defined by the second eigenvector (i.e., the second principal component).

Table 5.1 Eigenvectors Underlying the Correlation Matrix in Table 1.6 (and Corresponding to the Five Eigenvalues Presented in the Text Above)

Variable	Eigenvector 1	Eigenvector 2	Eigenvector 3	Eigenvector 4	Eigenvector 5
Ideal79	**.60**	**.11**	−.02	.58	.53
Expect79	**.62**	**−.19**	.28	.04	−.70
Number80	**−.19**	**.68**	−.28	.47	−.45
Ideal82	**.45**	**.31**	−.60	−.58	.04
Number04	**.08**	**.62**	.69	−.31	.16

Note: The first two eigenvectors, which we will use as the dominant eigenvectors, are in bold.

Figure 5.10   An Eigenvector Plot of the Correlation Matrix in Table 1.7, Representing NLSY Respondent Fertility Attitudes and Behaviors Between 1979 and 2004

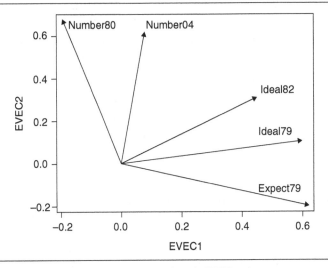

Note: NLSY = National Longitudinal Survey of Youth; EVEC = eigenvector.

We note two follow-up features of the eigenvector plot. First, Friendly (2002) developed the eigenvector plot at least in part to help order the variables. In many (most) circumstances, the order in which a researcher defines the variables is arbitrary. Friendly recognized that ordering similar variables together can help with the representation of a correlation matrix using many of the graphical procedures in this chapter. Based on the information from the eigenvector plot in Figure 5.10, we would be inclined to order the variables Expect79, Ideal79, Ideal82, Number04, and Number80 (which is slightly different from our original conceptual ordering). We note that there is a large technical literature on ordering variables in correlation matrices (and, more generally, stimuli in proximity matrices) in the psychometric literature; this method of ordering the elements of the matrix (variables, in the case of correlation matrices) based on internal features of the matrix is often referred to as "seriation." That literature is beyond (and beside) the scope of the current book, though the interested reader can consult Behrisch et al. (2016), Hubert (1974), or Rodgers and Thompson (1992) to gain access to the broader literature.

The second and final comment related to eigenvalue plots returns to the original discussion in the introduction to this section regarding the difference between graphical procedures of multiple bivariate patterns versus those that capture the underlying multivariate structure. The plot in Figure 5.10 clearly demonstrates that the eigenvalue plot falls into the second category. However, it is still incomplete in representing all the multivariate structure in a correlation matrix (unless the eigenvector plot is of all the variables, which is difficult to visualize beyond three variables in three dimensions). It is well beyond the human information-processing capacity to fully understand and appreciate—even using a sophisticated graphical tool to support the process—the complex multivariate interrelationships among many variables, as captured in a correlation matrix. The eigenvector plot—which does account for multivariate structure—is a good starting point for moving beyond the limitations of multiple bivariate representations.

Summary

Useful and insightful graphical methods have been developed to portray correlation matrices. As stated above regarding graphs of single correlations, a correlation matrix and a graph of the correlations in a correlation matrix portray different features of a set of multivariate relationships among several (or potentially many) variables. Studying both the correlation values and one or more graphical displays of a correlation matrix is strongly recommended.

We repeat—more broadly—some of the entreaties discussed in the conclusion of the last subsection. A careful study of the different correlation matrix graphs that are reviewed above will reveal that few of these graphs are sensitive to the multivariate structure that is naturally built into a correlation matrix beyond the many pairwise bivariate relationships (there are $p * (p - 1)/2$ of those relationships for p variables). This limitation may appear to be a disadvantage—and it is—but the disadvantage may not be fundamental on careful reflection. It is difficult for humans—even sophisticated researchers—to process and understand deep multivariate structure. That human limitation is exactly why many modeling methods applied to correlation matrices—factor analysis, PCA, SEM, and so on—such as those reviewed in Chapter 4 have been developed. Another way of stating this position is that the graphical portrayal of all pairwise correlations in a correlation matrix may push the limits of the human processing system already; to move beyond bivariate to multivariate relationships would likely, in at least some cases, push beyond the interpretable human limits. However, having stated this position, we note that the field of graphical analysis is wide open for the development of multivariate display methods (in addition to the eigenvector plot) that move beyond portraying many bivariate graphical relationships.

Chapter 6

THE GEOMETRY OF CORRELATION MATRICES

In the previous chapter, we described methods of graphing correlation matrices, as visualizing correlation matrices that are large or have a particular structure can provide more information for researchers or students than viewing the raw correlation matrix. In this chapter, we take the idea of visualizing correlation matrices further by discussing how to visualize the correlation space, which is a space that embeds correlation matrices within it. The material in this chapter may appear relatively abstract, and some of it is not highly relevant to the applied researcher. However, investigating the correlation space has led to recent discoveries regarding techniques involving correlation matrices, and learning about the correlation space can facilitate greater understanding of how correlation matrices are related to each other. Thus, this chapter is designed to focus on conceptual details and to be as accessible as the rest of the book. The first section of this chapter focuses on the definition and visualization of correlation space. Following, we discuss some known properties of correlation space. We conclude the chapter by discussing recent uses of and connections to correlation space in the literature.

What Is Correlation Space?

The correlation space is the visual space that all correlation matrices of a particular size occupy. Most useful correlation spaces—with a few important exceptions—apply to 3×3 correlation matrices. In the case of 3×3 correlation matrices, the correlation space can graphically show how different 3×3 true correlation matrices are related to each other and to pseudo-correlation matrices.

Each $p \times p$ correlation matrix is a single point in $p \times p$ correlation space. For example, Figure 6.1a demonstrates a 3×3 correlation matrix and Figure 6.1b shows how that correlation matrix can be graphed as a point in correlation space. The location of the point is denoted by a vector—that is, a list—of all the unique correlations in the matrix. Because of the symmetry and ones along the diagonal of correlation matrices, only the upper- or lower-triangular portion of the matrix need be represented; the number of unique correlations, q, where $q = [p(p - 1)]/2$, creates the ordered vector (generically termed r) that uniquely identifies the correlation matrix in correlation space. For example, consider the 4×4 correlation matrix (and note the symmetry and ones that define the diagonal).

$$\begin{bmatrix} 1 & -.23 & .04 & -.14 \\ -.23 & 1 & .35 & .05 \\ .04 & .35 & 1 & -.06 \\ -.14 & .05 & -.06 & 1 \end{bmatrix}$$

This correlation matrix can be uniquely identified by the vector $r = (-.23, .04, -.14, .35, .05, -.06)$, which has the $6 = (4 * 3)/2$ unique correlations ordered in the vector. We call r an ordered sextuple—"ordered" because the order of the correlations matters, and "sextuple" because there are six elements in the vector r. In general, each vector of unique correlations would be called an ordered q-tuple. Correlations can be ordered in r in any way desired as long as it is consistent; we chose to order the vector such that all correlations in the upper-triangular portion are arranged by stringing out the rows one after the other.

Note that p is the dimension of the correlation matrix (i.e., the number of variables), and q is the length of r as well as the dimension of the correlation space for correlation matrices of size p. Each axis of correlation space corresponds to a unique correlation in the correlation matrix; the x-axis refers to r_{12}, the y-axis refers to r_{13}, and so on. However, the axes do not

Figure 6.1   (a) A 3×3 Matrix and (b) the Associated Point It Occupies in a 3×3 Correlation Space

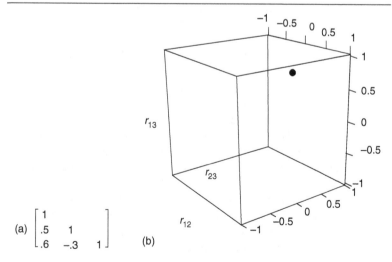

extend infinitely because correlation coefficients are bounded between [−1,1], and thus, the correlation space for $p \times p$ correlation matrices occupies a portion of the $[−1,1]^q$ hypercube—a cube in three dimensions (when $q = 3$), and a cube-like shape that exists in six or more dimensions when $q > 3$. The portion of the $[−1,1]^q$ hypercube that is not occupied by correlation space is where pseudo-correlation matrices live; therefore, the $[−1,1]^q$ hypercube can be separated into two parts: the part occupied by true correlation matrices and the part occupied by pseudo-correlation matrices.

Correlation space is easy to visualize for 3×3 correlation matrices. Each matrix is a single point inside a three-dimensional cube. Each of the three axes of the cube range from −1 to +1, and thus a 3×3 correlation matrix with elements $r_{12} = .3$, $r_{13} = .7$, $r_{23} = .5$—implying an r vector of $r = (.3, .7, .5)$—is the point inside this cube that lies at the coordinates $(.3, .7, .5)$. The point $(.2, −.4, −.3)$ corresponds to a different correlation matrix, and so on. All 3×3 correlation matrices (and all 3×3 pseudo-correlation matrices) lie inside the 3×3 correlation space. We define this space more generally and present geometric interpretations—pictures—in the next section.

The 3 × 3 Correlation Space

To visualize correlation space, consider a 3×3 true correlation matrix. The correlations among three variables, X_1, X_2, and X_3, produce a correlation matrix R of order $p = 3$, which can be represented in the ordered triple $r = (r_{12}, r_{13}, r_{23})$. The set of all possible rs (corresponding to true correlation matrices) within the cube $[−1, 1]^3$ is depicted in gray in Figure 6.2; this gray region is the 3×3 correlation space that was first conceptualized and defined by Rousseeuw and Molenberghs (1994).

The 3×3 correlation space has been described as an elliptical tetrahedron or an elliptope (Chai, 2014). The shape meets the edges of the cube at four of the eight "vertices" of the cube, corresponding to the most extreme 3×3 correlation matrices such that $r = (1,1,1)$, $r = (−1,−1,1)$, $r = (−1,1,−1)$, and r $= (1,−1,−1)$. These "extreme" correlation matrices represent perfect correlation (in the positive or negative direction) between all three pairs of variables. The four vertices of the correlation space show graphically that there are only four possible combinations of these perfect correlations. A correlation matrix where all three correlations are −1, that is, $r = (−1,−1,−1)$, would not be a true correlation matrix, for example. This r would correspond to a pseudo-correlation matrix (no real data exist that could produce this combination of correlations) and, therefore, falls outside the correlation space.

The portion of the 3×3 correlation space representing true correlation matrices includes diagonal lines across all six of the cube faces; these diagonal lines show graphically the mathematical necessity that if any one

88

Figure 6.2 The 3 × 3 Correlation Space

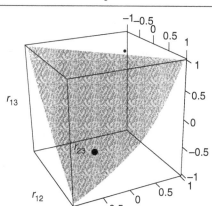

Note: The larger point ($r_{12} = .3, r_{13} = -.6, r_{23} = -.3$) lies within the correlation space and corresponds to a true correlation matrix R. The smaller point ($r_{12} = -.7, r_{13} = .8, r_{23} = .8$) lies outside the correlation space and corresponds to a pseudo-correlation matrix.

correlation of r_{12}, r_{23}, or r_{23} equals 1, the other two correlations must equal each other. This is a special case of the restriction of correlation range discussed in Chapter 2; in this case, the "range" reduces to a single possible value that the remaining two correlations must share. Similarly if any one correlation equals −1, the other two correlations must equal each other but be of opposite sign if the correlation matrix is to be PSD and therefore a true correlation matrix.

If any one correlation in a 3 × 3 correlation space is set equal to a value that is not 1 or −1, then the correlation space reduces to a two-dimensional plane of values the other two correlations can take on. Graphically, this involves taking slices of the cube parallel to an axis. (The "slicing" is determined by choosing the axis corresponding to the predetermined correlation and slicing at the value of the predetermined correlation.) Figure 6.3 demonstrates several slices of 3 × 3 correlation space in three and two dimensions. These correlation space slices, by setting the value of, say, r_{12}, would look like a diagonal line (at $r_{12} = -1$); then ellipses (between $r_{12} > -1$ and $r_{12} < 0$), which become more circular; then a perfect circle (at $r_{12} = 0$) before reverting back to ellipses (between $r_{12} > 0$ and $r_{12} < 1$); and finally to a diagonal line (at $r_{12} = 1$) in the opposite direction. Practically, this means that the possible values for two of the three correlations are the most constrained when the third correlation is an extreme correlation value (i.e., −1 or 1), and are the least constrained when the third correlation is (or is close to) 0.

Figure 6.3 Slices of 3 × 3 Correlation Space. (a) and (b) Two Different
Views of Three Slices of the Correlation Space: at $r_{23} = .3$
(Light Gray), $r_{23} = .6$ (Gray), and $r_{23} = .9$ (Black). (c) The
Same Slices in Two Dimensions

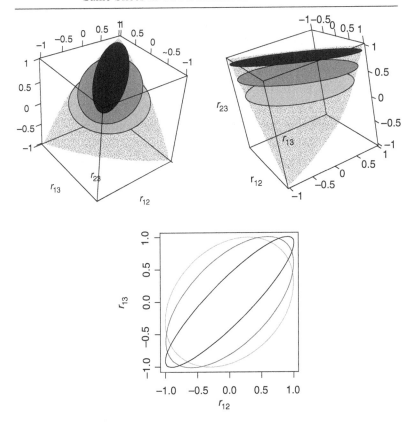

Note: As r_{23} gets larger, the range of values for r_{12} and r_{13} such that a true correlation matrix
is produced gets smaller.

In the case of 3 × 3 correlation matrices, the correlation space occupies
approximately 61.7% of the $[-1, 1]^3$ cube. Practically speaking, this means
that any random three correlations, generated uniformly between $[-1, 1]$ for
$r = (r_{12}, r_{13}, r_{23})$, which would correspond to a random 3 × 3 matrix of correlation coefficients, have a 61.7% chance of corresponding to a true
correlation matrix. Points in the cube that fall outside of the correlation
space correspond to pseudo-correlation matrices. Points on the surface of

the correlation space correspond to true correlation matrices which have at least one zero eigenvalue (i.e., a PSD, but not PD, correlation matrix, using the definitions presented in Chapter 2). An r close to the surface of the 3 × 3 correlation space corresponds to an R that has close-to-zero eigenvalues or, alternatively, has near-linear dependency among the three variables. Therefore, given two 3 × 3 correlation matrices, their corresponding r vector can be plotted in correlation space and readily compared visually in terms of their eigenvalues and linear dependence. The closer a correlation matrix's point is to the edge of the elliptote shown in Figure 6.2, the closer the eigenvalue(s) of the matrix is to zero.

Properties of Correlation Space: The Shape and Size

As soon as we consider larger correlation matrices (say, 4 × 4 correlation matrices), we can no longer visualize the entire correlation space as we can for 3 × 3 correlation matrices. Rousseeuw and Molenberghs (1994) presented visual examples of "banded" 4 × 4 correlation matrices, though these are very special cases. However, even if we can't see these 4 × 4 spaces with our limited three-dimensional vision, we know these spaces have certain geometric and mathematical properties.

Convexity of the Space

The correlation space for $p \times p$ correlation matrices—plotted in a $q = p$ $(p-1)/2$-hypercube—is convex (Chai, 2014). Convexity is a mathematical property that means a line drawn between any two points in $q \times q$ correlation space will be fully contained within the correlation space. Practically, this means that any weighted average between two true correlation matrices of the same size will itself be a true correlation matrix. However, this property only applies to correlation matrices of the same size, as each correlation space (of size q) is convex, but there is no guarantee of convexity across correlation spaces for different sizes of q. This is why averaging true correlation matrices of different orders does not guarantee that the resultant matrix is a true correlation matrix.

Number of Vertices and Edges

"Vertices" in correlation spaces correspond to true correlation matrices that contain complete redundancy and perfect correlation among the variables in the correlation matrix. These extreme correlation matrices have off-diagonal elements that each is either −1 or 1, and the eigenvalues of these matrices are such that the largest eigenvalue is equal to p and all other eigenvalues are 0. For example, the four vertices in 3 × 3 correlation space

Table 6.1 Number of Vertices of Correlation Spaces in q Dimensions for Correlation Matrices of Order p

p	q	No. of Vertices (q-Hypercube)	No. of Vertices (Correlation Space)
1	0	1	1
2	1	2	2
3	3	8	4
4	6	64	8
5	10	1,024	16
⋮	⋮	⋮	⋮
p	$q = p(p-1)/2$	2^q	2^{p-1}

listed above each correspond to correlation matrices with the largest eigenvalue equal to 3 and two eigenvalues equal to 0. Extreme correlation matrices that correspond to vertices in correlation space are obviously rare in practice but are useful for imagining the boundary of the correlation space for correlations larger than order 3; even though we cannot visualize these spaces, we know certain information about the number of vertices of correlation spaces, shown in Table 6.1.

Furthermore, the space of correlation matrices in higher dimensions must have a diagonal edge along each hypercube face, corresponding to the degenerate cases of perfect (negative or positive) association among variables. The face of a hypercube corresponds to possible values of two correlations (say, r_{xw} and r_{xv}) after all of the other correlations in the matrix have been defined as either −1 or 1. Given the specified values of all the other correlations, r_{xw} will equal r_{xv} or $−r_{xv}$ for the correlation matrix to be PSD, depending on the pattern in the correlation matrix that produces the diagonal line observed in the hypercube.

Volume Relative to Space of Pseudo-Correlations

In the 3 × 3 case, approximately 61.7% of the [−1, 1] cube is occupied by the correlation space. However, as p increases, the number of vertices of correlation space relative to the vertices of the hypercube it occupies decreases; similarly, the proportion of the hypercube occupied by correlation space also decreases. Böhm and Hornik (2014) derived the analytic proportions of the hypercubes that correlation spaces occupy (see Table 6.2). As shown, the proportions decrease exponentially, such that the correlation space for 6 × 6 correlation matrices occupies less than 0.1% of the [−1,1][15]

92

Table 6.2 The Proportion of the Hypercube Occupied by Correlation Spaces

p	Proportion of Hypercube
2	1.00000000
3	0.61685027
4	0.18277045
5	0.02200445
6	0.00094952
7	0.00001328

Note: Reproduced from Böhm and Hornik (2014).

hypercube. Practically, this means that a 6 × 6 matrix with uniform randomly generated values between [−1, 1] as the off-diagonal elements (and 1s along the diagonal) has less than a 0.1% chance of being a PSD, therefore corresponding to a true correlation matrix. This sharp decrease in the proportion of correlation space has led researchers to develop many different ways to generate random, true correlation matrices for simulation research.

Uses of Correlation Space

Similarity of Correlation Matrices

In Chapter 3, we covered how to statistically test the hypothesis that two correlation matrices were equal in the population. However, there is a broader treatment for assessing the similarity—rather than the strict equality—of matrices generally, and correlation matrices specifically. There are many measures of matrix "norms," or distances, in the linear algebra literature to capture the many ways that mathematicians and statisticians may wish to measure distance or similarity of matrices. Several of these matrix norms can be visualized within the bounds of correlation space.

The easiest method of visualizing the distance between correlation matrices is the Euclidean distance between them in correlation space. Euclidean distance, also called straight-line distance, is the standard distance formula used in everyday distance calculations if we were to calculate the distance between two points "as the crow flies." Euclidean distance between two correlation matrices of the same order is calculated by squaring the difference between corresponding correlations, summing these differences, and taking the square root of the product. Geometrically, the Euclidean distance between two matrices is the length of the straight line between them in correlation space (see Figure 6.4). We could use Euclidean distance to measure

Figure 6.4   Euclidean Distance in 3 × 3 Correlation Space

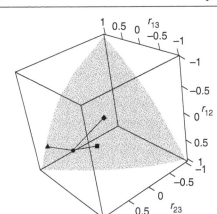

Note: The Euclidean distance from one correlation matrix $r = (.5, .3, .6)$ (circle) to another correlation matrix $r = (.4, -.1, .3)$ (square), to the identity matrix $r = (0, 0, 0)$ (diamond), and to a correlation matrix on the boundary of the correlation space $r = (.35, .95, .60)$ (triangle).

the distance between two correlation matrices, a correlation matrix and the corresponding identity correlation matrix (i.e., a correlation matrix that shows independence among variables), or the distance of a correlation matrix from the boundary of the correlation space. Because of the "convexity" property of the correlation space, discussed above, the entire straight line showing the Euclidean distance between two matrices is always itself within correlation space.

However, there are cases where the Euclidean distance may not be preferred. This is often the case in day-to-day life, such as getting directions to a new location: Being provided the "as the crow flies" distance to a new coffee shop is not as useful as the distance along streets, which may be a much longer distance than the Euclidean distance. Because Euclidean distance calculates squared differences between correlations, exceptionally large differences between two correlations may disproportionately influence the calculated distance. Using absolute values of differences rather than squares of differences may be a preferred method of calculating distance in these cases. This type of distance formula is sometimes called Manhattan distance or taxicab distance, as it corresponds to the distance between two points as if tracing the distance between the points on a grid (much like maps of city blocks in a well laid-out city; Krause, 1986). On the other hand, instances may call for measuring distance by the maximum difference between corresponding correlations (a dominance distance), or by a subset of correlations of interest. The Manhattan distance, the Euclidean distance, and the dominance distance are all special cases of what measurement specialists refer to as the Minkowski metric.

Depending on the selected measure of distance between two correlation matrices, a researcher may draw differing conclusions as to how similar two correlation matrices are. The variety of measures of distance have also informed other related fields of research for correlation matrices, including methods of measuring the equality of correlation matrices, correlation matrix smoothing techniques (described in Chapter 2), and methods of generating correlation matrices similar to a target correlation matrix. All of these fields of research are interpreted within the context of correlation space.

Generating Random Correlation Matrices

Monte Carlo studies are studies that use computer simulations to produce many data replications under controlled conditions to test the performance of quantitative techniques across a variety of data conditions. Monte Carlo studies in psychology, economics, genetics, and other areas that use quantitative methods may require many randomly generated correlation matrices from across correlation space to capture a wide range of potential correlation matrix structures that might be observed in real data (Hardin et al., 2013). However, because of the PSD constraint on correlation matrices, the task of generating random correlation matrices is nontrivial. The simplest method to generate correlation matrices is called the rejection method, which involves generating a random number between $[-1, 1]$ for each off-diagonal element of the matrix and "rejecting" the matrix if it is not PSD—that is, if the generated matrix is a pseudo-correlation matrix. The rejection method quickly becomes infeasible as p becomes large because the ratio of non-PSD pseudo-correlation matrices to PSD true-correlation matrices gets large quickly, as shown in Table 6.2 (Böhm & Hornik, 2014; Numpacharoen & Atsawarungruangkit, 2012). This has led to several novel methods of generating random correlation matrices across correlation space—based on partial correlations (Joe, 2006), eigenvalues (e.g., Chalmers, 1975), expected means (Marsaglia & Olkin, 1984), or factor structure (e.g., Tucker et al., 1969). Each method seeks to sample correlation matrices randomly across correlation space, but each samples matrices in different ways and with different distributions of correlations.

Fungible Correlation Matrices

In the ordinary least squares (OLS) regression framework, fungible correlation matrices refer to a set of possible correlation matrices among the predictor (X) variables that is different from the observed correlation matrix among the predictor variables (Waller, 2016). For a particular Y and set of X variables, OLS produces a unique, maximally large squared multiple correlation (SMC) and a unique set of OLS regression coefficient estimates. Waller (2016) demonstrates that an infinite number of correlation matrices among the X variables will produce these same SMC and regression coefficient estimates; he refers to this infinite set of correlation matrices as "fungible," or exchangeable.

Fungible correlation matrices can be strictly true-correlation matrices, strictly pseudo-correlation matrices, or both. Geometrically, the set of fungible correlation matrices is defined by slices of correlation space that are not necessarily perpendicular to any axis. These slices are planes in three dimensions and hyperplanes in higher dimensions; the orientation of the plane is defined by the regression coefficients and SMC. Furthermore, these slices do not create ellipses in correlation space (as demonstrated in Figure 6.3) but rather rounded triangles—Waller (2016) likened such intersections to guitar picks.

Waller (2016) showed how the set of fungible correlation matrices could be used in Monte Carlo research, specifically to compare alternative regression techniques with OLS (in this case, penalized regression techniques) and smoothing techniques (some of which were reviewed in Chapter 2). Fungible correlation matrices have properties different from other correlation matrix generation methods described above because they are developed by slicing the correlation space, allowing researchers more control over the types of correlation matrices generated. Code for generating fungible correlation matrices is provided in Jones and Waller (2016) and Waller (2016).

Defining Correlation Spaces Using Angles

The correlation space defined by Rousseeuw and Molenberghs (1994) defined correlations that ranged from −1 to +1 in a [−1, 1] hypercube and produces a 3 × 3 correlation space that is an elliptope. Another natural interpretation of the correlation is as an angle (between 0° and 180°) between variable vectors (see Chapter 5). This angular interpretation leads to the suggestion (see Chai, 2014; Hadd, 2016) that the cube to define correlation space should range from [0°, 180°] (whose cosines range from −1 to +1, as they must for correlations; see Chapter 5). When the correlation space is graphed using angles as units, an interesting and parsimonious structure emerges. The graph of 3 × 3 correlation space becomes a perfect tetrahedron, as seen in Figure 6.5, rather than the elliptical tetrahedron in Figure 6.2. Though Chai (2014) was the first to describe this tetrahedron verbally and mathematically, the authors of this book discovered the structure independently at approximately the same time. The first visual/graphical

Figure 6.5   The Structure of Correlation Space When Axes Are Defined as Angles Between 0° and 180°

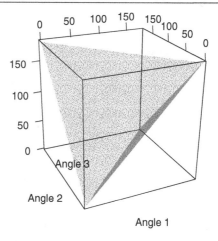

Angle 1

presentation of the tetrahedron in Figure 6.5 was in a conference presentation by Rodgers and Hadd (2015) and in Hadd's (2016) master's thesis, and Figure 6.5 is the first time of which we are aware that this visual interpretation has been presented in the statistical literature. Using Figure 6.5 in introductory and more advanced statistics can facilitate appreciation of the relationships among correlation matrices within correlation space, and also the relation between true- and pseudo-correlation matrices.

It is easy to prove that 3×3 tetrahedrons occupy exactly 1/3 of the cube in which they are inscribed. Thus, correlation space defined in relation to angles occupies 33% instead of 61.7% of the total space (see Table 6.2). Furthermore, Hadd (2016) showed that correlation spaces using angles for higher than 3×3 correlation matrices are not hypertetrahedra, they are rather more complex geometric structures, a special case of which in the 3×3 case is a tetrahedron. Generating random correlation matrices in this angle-based hypercube may ultimately have advantages over generating them in the space originally defined by Rousseeuw and Molenberghs (1994) and can help generate distributions of correlation matrices with properties different from other correlation generation methods.

Example Using 3×3 and 4×4 Correlation Space

We now present an example that illustrates how correlation space may, to some extent, be visualized for higher-dimension correlation matrices. We cannot directly visualize correlation spaces for 4×4 correlation spaces and

Figure 6.6   The Sports and Racial Composition Correlation Matrices in
Correlation Space

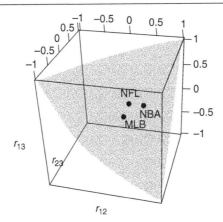

Note: The fourth variable of each matrix has been removed. The points corresponding to the
MLB, NBA, and NFL correlation matrices are labeled. MLB = Major League Baseball;
NBA = National Basketball Association; NFL = National Football League.

larger, so we cannot directly show any of our example correlation matrices
in correlation space. However, using the sport and racial composition correlation matrices in Table 1.5, we can show an adjusted 3 × 3 correlation
space, whereby the last variable (% Black '90) is dropped from each correlation matrix. Now each correlation matrix has three correlations, rather
than the six that are actually shown in Table 1.5. Figure 6.6 shows the
points in space that correspond to the NBA, NFL, and MLB correlation
matrices, respectively. Notice that by Euclidean distance, the MLB and the
NBA correlation matrices appear furthest apart from each other.

   However, the correlation space in Figure 6.6 assumes that the three omit-
ted correlations (r_{14}, r_{24}, r_{34}) in each matrix—those associated with the "%
Black '90" variable"—can be *any* possible correlations, rather than the
predetermined correlations that they actually are. If we constrain the 4 × 4
correlation space such that $r_{14} = -.06$, $r_{24} = .29$, and $r_{34} = .99$ for NBA, $r_{14} =$
$-.10$, $r_{24} = .30$, and $r_{34} = .99$ for NFL, and $r_{14} = -.23$, $r_{24} = .04$, and $r_{34} = .96$
for MLB, then the resulting 3 × 3 space that the remaining correlations can
occupy will look very different from the elliptical tetrahedron in Figure 6.2.
These new spaces, shown in Figure 6.7, can be considered "slices" of the
higher dimension 6 × 6 space that we cannot see, with slices at different
values of the r_{14}, r_{24}, and r_{34} axes generating different 3 × 3 correlation

98

Figure 6.7   Slices of 4 × 4 Correlation Space Associated With the (a) NBA, (b) NFL, and (c) MLB Correlation Matrices

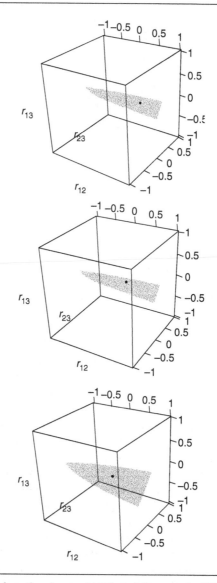

Note: The gray area shows the subspace of 4 × 4 correlation space that PSD correlation matrices may occupy once the values of the r_{14}, r_{24}, and r_{34} are predetermined. The spaces differ because the values of the r_{14}, r_{24}, and r_{34} differ (see Table 1.5). The black point in each space is associated with the observed NBA, NFL, and MLB correlation matrices, respectively. NBA = National Basketball Association; NFL = National Football League; MLB = Major League Baseball.

subspaces. Notice that the subspace for the MLB correlation matrices (Figure 6.7c) is larger and less constrained than the other two subspaces; this indicates that the combination of values of r_{14}, r_{24}, and r_{34} for the MLB correlation matrix provides more options for the other correlations in the matrix to produce a PSD correlation matrix compared with the NFL and NBA correlation matrices. This insight is difficult to discuss, and nearly impossible to visualize, without the use of correlation space.

Summary

The current chapter is more abstract and mathematical than the previous five, but we believe that the concept of correlation space is a powerful conceptual tool to allow researchers and students to better understand both specific correlation matrices and the relationships among two or more correlation matrices. Furthermore, the role of eigenvalues in defining correlation matrices is expanded and clarified within correlation space. In fact, we believe that any time a geometric/visual interpretation can be added to a set of algebraic concepts, expanded and clarified interpretation is likely to emerge.

Chapter 7

CONCLUSION

John B. Carroll (1961), in his Psychometric Society presidential address, stated that the "correlation coefficient is one of the most frequently used tools of psychometricians . . . and perhaps also one of the most frequently misused" (p. 347). In the almost 60 years since that statement, statistical analysis has been trending in the direction of becoming evermore multivariate (for support, see, among many others, Harlow, 2005). In 2020, when this concluding chapter is being written, it is safe to rephrase Carroll's statement: The correlation matrix is one of the most frequently used tools of both methodologists and researchers across many different disciplines. To support understanding the correlation matrix, and to avoid misuse of this important statistical tool, has been the goal of this book.

A single correlation coefficient measures the relationship between two variables; a correlation matrix displays the relationships among many variables. The shift from a bivariate relationship to multivariate relationships has many and deep implications, at both the mathematical and the substantive level. Substantively, multivariate analysis allows a researcher to study many variables simultaneously; to control many variables, as the researcher focuses on the relationships among a smaller subset of important variables; and to even shift the importance of variables back and forth within the same model. It is not unusual for a variable, X_1, to serve to control for confounds between variables X_2 and X_3 in one part of a model, to serve along with X_4 as the predictor of X_5 in another part of the model, and to be the dependent variable predicted from X_6 and X_7 in another part of the model. Only a multivariate perspective allows such flexibility in the evaluation of behavioral models.

Underlying those potentially valuable substantive models and interpretations is a set of multivariate data spaces, ones that are mathematically elegant and facile in capturing the substantive processes described briefly in the previous paragraph. The two primary multivariate data spaces each emerge from the same N observation by p variable data set (with, e.g., $N = 100$ seventh graders as rows, and scores for those students on $p = 4$ variables—height, weight, jumping ability, and running speed—defining the four columns). The two kinds of spaces are the several potential variable spaces, in which scatterplots are defined as N points, each of which is an ordered p-tuple (or fewer) of scores; and several subject spaces, in which p (or fewer) variable vectors define angles that reflect the correlations among p (or fewer) variables lying in N-dimensional (or possibly smaller) space. The variable space is often taught and is the basis for scatterplots and scatterplot matrices. The second multivariate data space, the subject space, is an underutilized though highly valuable interpretational tool as well. It is the space in which correlations are related to angles. This second space can provide overlapping information compared with the variable space, but subject space can also render new and unique interpretations of the data. Both spaces will facilitate researchers in understanding the patterns underlying their multiple variables.

There have been several important themes that a careful reader will notice running through this book, which we summarize in this conclusion. First, a correlation matrix should be viewed as rather more nuanced and valuable than simply as an organized collection of bivariate correlations (see Chapters 1 and 2 in particular, and other chapters have continued to address this theme). The set of bivariate correlations is useful in its own right, for example, as the basis for a scatterplot matrix or as measures of the relationship among all pairs of variables. However, the additional nuance comes from the multivariate constraints built into a correlation matrix. Not only are the correlations between X_1 and X_2, between X_1 and X_3, and between X_2 and X_3 of interest to a substantive researcher, but each of those correlations places constraints on the other two that reflect the *multivariate* relationships among the three variables. The more variables, the more complex and nuanced are those constraints.

One way to conceptualize this added nuance is to consider that among, for example, five variables, there are 10 unique pairwise relationships, 6 three-way relationships, 5 four-way relationships, and 1 five-way relationship. A strictly bivariate interpretation would focus on the 10 pairwise relationships but would miss the 12 additional higher-way relationships. We hasten to add, however, that the human information-processing system is limited in its ability to interpret such highly multivariate relationships. That is a human limitation, however, and not a mathematical or statistical limitation.

A second theme that has been running through this book is that the correlation matrix provides diagnostic indicators of various features of the matrix (see Chapter 2). Among the most useful of those are the eigenvalues of a correlation matrix, which indicate whether the correlation matrix is a true correlation matrix or a pseudo-correlation matrix (which is a collection of correlations arrayed in a matrix, but one for which no possible set of variables exists that could have generated those correlations). As we noted in Chapter 6, the more variables, the higher the fraction of the correlation space that is occupied by pseudo-correlation matrices, and the smaller the fraction of the correlation space occupied by true correlation matrices.

A third theme is that various statistical procedures have been developed to study and evaluate correlation matrices (see Chapter 3). These procedures are, by definition, multivariate statistical procedures, and therefore require multivariate analysis using linear (matrix) algebra to obtain statistical results. Thousands of research articles have been published in a broad interdisciplinary scientific literature in which statistical analysis of patterns in correlation matrices was prerequisite to successful evaluation of one or more research hypotheses.

A fourth theme is that correlation matrices have become the raw data for a number of sophisticated quantitative modeling methods (see Chapter 4). Examples include procedures such as factor and components analysis, SEM, multilevel modeling, categorical data analysis, hazards modeling, mixture modeling, growth curve modeling, and meta-analysis (this list is nowhere close to exhaustive). These statistical modeling methods require correlation matrices (or, often, the unstandardized version of a correlation matrix, a covariance matrix) as the basic input to the modeling method.

A fifth theme is that many graphical methods have been developed to support applied researchers in studying and understanding the bivariate and multivariate patterns built into correlation matrices (see Chapter 5). We argued, and demonstrated, how necessary it is for researchers to study *both* the correlations themselves and also the graphical patterns underlying the correlations. Many graphical methods have been developed since Tukey decried the limitations of graphical software for demonstrating patterns of value to statisticians and researchers. Loosely, these can be broken down into those that graph pairs of variables—bivariate methods, which apply to all of the bivariate correlations contained in a correlation matrix; and multivariate methods, which apply to some or all of the higher-way relationships defined in a correlation matrix. At this writing, most graphical methods useful in portraying correlation matrices fall into the former category. Developing new and innovative ways to visually portray the multivariate patterns—as the eigenvector plot does—deserves attention from researchers who study and develop graphical methods.

Finally, we also demonstrated some potential areas of recent growth, and potential future research arenas, that are relevant to understanding correlation matrices (see Chapter 6). These rely heavily on the concept of the correlation space, which was defined carefully and demonstrated both conceptually and mathematically. Geometric interpretations of 3×3 correlation space are particularly valuable as a pedagogical and interpretational tool.

The correlation coefficient was first defined by Pearson (1895), 125 years ago. The correlation matrix was then developed and used in a substantive psychological research study on human intelligence by Spearman (1904), 116 years ago. We understand a great deal about correlation matrices, as this book has demonstrated. Perhaps, however, the best is yet to come, especially at substantive levels in which correlation matrices are used in applied behavioral research settings by well-trained researchers who naturally think of the world through a multivariate lens.

REFERENCES

Acock, A. C. (2013). *Discovering structural equation modeling using Stata.* Stata Press.

Anderson, T. W. (1963). Asymptotic theory for principal component analysis. *Annals of Mathematical Statistics, 34*(1), 122–148. https://doi.org/10.1214/aoms/1177704248

Anscombe, F. J. (1973). Graphs in statistical analysis. *The American Statistician, 27*(1), 17–21. https://doi.org/10.1080/00031305.1973.10478966

Bandalos, D. L. (2017). *Measurement theory and applications for the social sciences.* Guilford Press.

Beaujean, A. A. (2014). *Latent variable modeling using R: A step-by-step guide.* Routledge/Taylor & Francis. https://doi.org/10.4324/9781315869780

Behrisch, M., Bach, B., Riche, N. H., Schreck, T., & Fekete, J.-D. (2016). Matrix reordering methods for table and network visualization. *Computer Graphics, 35*, 693–716. https://doi.org/10.1111/cgf.12935

Bentler, P. M., & Bonett, D. G. (1980). Significance tests and goodness of fit in the analysis of covariance structures. *Psychological Bulletin, 88*(3), 588–606. https://doi.org/10.1037/0033-2909.88.3.588

Bentler, P. M., & Yuan, K. H. (1998). Tests for linear trend in the smallest eigenvalues of the correlation matrix. *Psychometrika, 63*, 131–144. https://doi.org/10.1007/BF02294771

Bentler, P. M., & Yuan, K. H. (2011). Positive definiteness via off-diagonal scaling of a symmetric indefinite matrix. *Psychometrika, 76*(1), 119–123. https://doi.org/10.1007/s11336-010-9191-3

Böhm, W., & Hornik, K. (2014). Generating random correlation matrices by the simple rejection method: Why it does not work. *Statistics & Probability Letters, 87*, 27–30. https://doi.org/10.1016/j.spl.2013.12.012

Bollen, K. A. (1989). *Structural equations with latent variables.* Wiley. https://doi.org/10.1002/9781118619179

Box, J. F. (1978). *R. A. Fisher: The life of a scientist*. Wiley.

Brown, T. A. (2015). *Confirmatory factor analysis for applied research*. Guilford Press.

Byrne, B. M. (2016). *Structural equation modeling with AMOS*. Routledge. https://doi.org/10.4324/9781315757421

Carroll, J. B. (1961). The nature of the data, or how to choose a correlation coefficient. *Psychometrika, 26*, 347–372. https://doi.org/10.1007/BF02289768

Chai, K. M. A. (2014). Three-by-three correlation matrices: Its exact shape and a family of distributions. *Linear Algebra and Its Applications, 458*, 589–604. https://doi.org/10.1016/j.laa.2014.06.039

Chalmers, C. P. (1975). Generation of correlation matrices with a given eigen-structure. *Journal of Statistical Computation and Simulation, 4*(2), 133–139. https://doi.org/10.1080/00949657508810116

Chambers, J. M., Cleveland, W. S., Kleiner, B., & Tukey, P. A. (1983). *Graphical methods for data analysis*. Chapman & Hall.

Chen, P. Y., & Popovich, P. M. (2002). *Correlation: Parametric and nonparametric measures* (Quantitative Applications in the Social Sciences No. 139). Sage. https://doi.org/10.4135/9781412983808.n1

Cheung, M. W.-L. (2014). Fixed- and random-effects meta-analytic structural equation modeling: Examples and analyses in R. *Behavior Research Methods, 46*(1), 29–40. https://doi.org/10.3758/s13428-013-0361-y

Cheung, M. W.-L., & Chan, W. (2005). Meta-analytic structural equation modeling: A two-stage approach. *Psychological Methods, 10*(1), 40–64. https://doi.org/10.1037/1082-989X.10.1.40

Cleveland, W. S. (1979). Robust locally weighted regression and smoothing scatterplots. *Journal of the American Statistical Association, 74*, 829–836. https://doi.org/10.1080/01621459.1979.10481038

Cleveland, W. S. (1993). *Visualizing data*. Hobart Press.

Dawson, K. S., Gennings, C., & Carter, W. H. (1997). Two graphical techniques useful in detecting correlation structure in repeated measures data. *The American Statistician, 51*, 275–283. https://doi.org/10.1080/00031305.1997.10473981

Dunn, O. J., & Clark, V. (1969). Correlation coefficients measured on the same individuals. *Journal of the American Statistical Association, 64*(325), 366–377. https://doi.org/10.1080/01621459.1969.10500981

Dunteman, G. H. (1989). *Principal components analysis* (Quantitative Applications in the Social Sciences No. 69). Sage. https://doi.org/10.4135/9781412985475

Elgar, F. J. (2010). Income inequality, trust, and population health in 33 countries. *American Journal of Public Health, 100*(11), 2311–2315. https://doi.org/10.2105/AJPH.2009.189134

Finch, W. H. (2019). *Exploratory factor analysis* (Quantitative Applications in the Social Sciences No. 182). Sage.

Friendly, M. (2002). CORRGRAMS: Exploratory displays for correlation matrices. *The American Statistician, 56*(4), 316–324. https://doi.org/10.1198/000313002533

Galton, F. (1885). Regression towards mediocrity in heredity stature. *Journal of the Anthropological Institute, 15*, 246–263. https://doi.org/10.2307/2841583

Hadd, A. R. (2016). *Correlation matrices in cosine space* [Unpublished master's thesis]. Vanderbilt University.

Hardin, J., Garcia, S. R., & Golan, D. (2013). A method for generating realistic correlation matrices. *Annals of Applied Statistics, 7*(3), 1733–1762. https://doi.org/10.1214/13-AOAS638

Harlow, L. L. (2005). *The essence of multivariate thinking: Basic themes and methods.* Psychology Press. https://doi.org/10.4324/9781410612687

Higham, N. J. (2002). Computing the nearest correlation matrix: A problem from finance. *IMA Journal of Numerical Analysis, 22*(3), 329–343. https://doi.org/10.1093/imanum/22.3.329

Hubert, L. J. (1972). A note on the restriction of range for Pearson product-moment correlation coefficients. *Educational and Psychological Measurement, 32*(3), 767–770. https://doi.org/10.1177/001316447203200315

Hubert, L. J. (1974). Some applications of graph theory and related non-metric techniques to problems of approximate seriation: The case for symmetric proximity measures. *British Journal of Mathematical and Statistical Psychology, 27*, 133–153. https://doi.org/10.1111/j.2044-8317.1974.tb00534.x

Humphreys, L. G., Davey, T. C., & Park, R. K. (1985). Longitudinal correlation analysis of standing height and intelligence. *Child Development, 56*(6), 1465–1478. https://doi.org/10.2307/1130466

Inselberg, A. (1985). The plane with parallel coordinates. *The Visual Computer, 1*(2), 69–91. https://doi.org/10.1007/BF01898350

Jacoby, W. G. (1997). *Statistical graphics for univariate and bivariate data* (Quantitative Applications in the Social Sciences No. 117). Sage. https://doi.org/10.4135/9781412985963

Jennrich, R. I. (1970). An asymptotic χ^2 test for the equality of two correlation matrices. *Journal of the American Statistical Association, 65*(330), 904–912. https://doi.org/10.1080/01621459.1970.10481133

Joe, H. (2006). Generating random correlation matrices based on partial correlations. *Journal of Multivariate Analysis, 97*(10), 2177–2189. https://doi.org/10.1016/j.jmva.2005.05.010

Jones, J. A., & Waller, N. G. (2016). Fungible weights in logistic regression. *Psychological Methods, 21*(2), 241. https://doi.org/10.1037/met0000060

Jöreskog, K. G. (1970). A general method for analysis of covariance structures. *Biometrika, 57*(2), 239–251. https://doi.org/10.1093/biomet/57.2.239

Kelloway, E. K. (2014). *Using Mplus for structural equation modeling.* Sage.

Kim, J. O., & Mueller, C. W. (1978a). *Factor analysis: Statistical methods and practical issues* (Quantitative Applications in the Social Sciences No. 14). Sage.

Kim, J. O., & Mueller, C. W. (1978b). *Introduction to factor analysis: What it is and how to do it* (Quantitative Applications in the Social Sciences No. 13). Sage.

Kline, R. B. (2015). *Principles and practice of structural equation modeling* (4th ed.). Guilford Press.

Kracht, J., & Waller, N. (2018). A comparison of matrix smoothing algorithms. *Multivariate Behavioral Research, 53*(1), 136–137. https://doi.org/10.1080/00273171.2017.1404899

Krause, E. F. (1986). *Taxicab geometry: An adventure in non-Euclidean geometry.* Dover.

Kullback, S. (1967). On testing correlation matrices. *Applied Statistics, 16*(1), 80–85. https://doi.org/10.2307/2985240

Leonard, W. M. (1997). Racial composition of NBA, NFL, and MLB teams and racial composition of franchise cities. *Journal of Sport Behavior, 20*(4), 424–434.

Lewis-Beck, C., & Lewis-Beck, M. (1980). *Applied regression: An introduction* (Quantitative Applications in the Social Sciences No. 22). Sage. https://doi.org/10.4135/9781412983440

Loh, W.-Y. (1987). Does the correlation coefficient really measure the degree of clustering around the line? *Journal of Educational and Behavioral Statistics, 12*(3), 235–239. https://doi.org/10.3102/10769986012003235

Long, J. S. (1983). *Covariance structure models: An introduction to LISREL* (Quantitative Applications in the Social Sciences No. 34). Sage. https://doi.org/10.4135/9781412983822

Marsaglia, G., & Olkin, I. (1984). Generating correlation matrices. *SIAM Journal on Scientific and Statistical Computing, 5*(2), 470–475. https://doi.org/10.1137/0905034

Meng, X. L., Rosenthal, R., & Rubin, D. B. (1992). Comparing correlated correlation coefficients. *Psychological Bulletin, 111*(1), 172. https://doi.org/10.1037/0033-2909.111.1.172

Murdoch, D. J., & Chow, E. D. (1996). A graphical display of large correlation matrices. *The American Statistician, 50*, 178–180. https://doi.org/10.1080/00031305.1996.10474371

Numpacharoen, K., & Atsawarungruangkit, A. (2012). Generating correlation matrices based on the boundaries of their coefficients. *PLOS ONE, 7*(11), e48902. https://doi.org/10.1371/journal.pone.0048902

Pearson, K. (1895). Notes on regression and inheritance in the case of two parents. *Proceedings of the Royal Society of London, 58*, 240–242. https://doi.org/10.1098/rspl.1895.0041

Pearson, K. (1896). Mathematical contributions to the theory of evolution: III—Regression, heredity and panmixia. *Philosophical Transactions of the Royal Society of London, 187*, 253–318. https://doi.org/10.1098/rsta.1896.0007

Pearson, K. (1901). On lines and planes of closest fit to systems of point in space. *The Philosophical Magazine, 2*(11), 559–572. https://doi.org/10.1080/14786440109462720

Preacher, K. J., Wichman, A. L., MacCallum, R. C., & Briggs, N. C. (2008). *Latent growth curve modeling* (Quantitative Applications in the Social Sciences No. 157). Sage. https://doi.org/10.4135/9781412984737

Protogerou, C., Johnson, B. T., & Hagger, M. S. (2018). An integrated model of condom use in sub-Saharan African youth: A meta-analysis. *Health Psychology, 37*(6), 586–602. https://doi.org/10.1037/hea0000604

Quinn, J. M., & Wagner, R. K. (2018). Using meta-analytic structural equation modeling to study developmental change in relations between language and literacy. *Child Development, 89*(6), 1956–1969. https://doi.org/10.1111/cdev.13049

Raveh, A. (1985). On the use of the inverse of the correlation matrix in multivariate data analysis. *The American Statistician, 39*(1), 39–42. https://doi.org/10.1080/00031305.1985.10479384

Rodgers, J. L. (2000). Social contagion and adolescent sexual behavior: Theoretical and policy implications. In J. Bancroft (Ed.), *The role of theory in sex research* (pp. 258–278). Kinsey Institute.

Rodgers, J. L. (2010). The epistemology of mathematical and statistical modeling: A quiet revolution. *American Psychologist, 65*(1), 1–12. https://doi.org/10.1037/a0018326

Rodgers, J. L. (2019). Degrees of freedom at the start of the second 100 years: A pedagogical treatise. *Advances in Methods and Practices in the Psychological Sciences, 2*, 396–405. https://doi.org/10.1177/2515245919882050

Rodgers, J. L., & Hadd, A. R. (2015, October). *Transforming correlation space using cosines versus linear intervals: Tetrahedra and simplex structures* [Paper presentation]. Meeting of the Society of Multivariate Experimental Psychology, Redondo Beach, CA, USA.

Rodgers, J. L., & Nicewander, W. A. (1988). Thirteen ways to look at the correlation coefficient. *The American Statistician, 42*(1), 59–66. https://doi.org/10.1080/00031305.1988.10475524

Rodgers, J. L., & Thompson, T. D. (1992). Seriation and multidimensional scaling: A data analysis approach to scaling asymmetric proximity matrices. *Applied Psychological Measurement, 16*(2), 95–117. https://doi.org/10.1177/014662169201600201

Rousseeuw, P. J., & Molenberghs, G. (1993). Transformation of non positive semidefinite correlation matrices. *Communications in Statistics, Theory and Methods, 22*(4), 965–984. https://doi.org/10.1080/03610928308831068

Rousseeuw, P. J., & Molenberghs, G. (1994). The shape of correlation matrices. *The American Statistician, 48*(4), 276–279. https://doi.org/10.1080/00031305.1994.10476079

Spearman, C. (1904). "General intelligence," objectively determined and measured. *American Journal of Psychology, 15*(2), 201–293. https://doi.org/10.2307/1412107

Stanley, J. C., & Wang, M. D. (1969). Restrictions on the possible values of $r12$ given $r13$ and $r23$. *Educational and Psychological Measurement, 29*(3), 579–581. https://doi.org/10.1177/001316446902900304

Stanton, J. M. (2001). Galton, Pearson, and the peas: A brief history of linear regression for statistics instructors. *Journal of Statistics Education, 9*(3). https://doi.org/10.1080/10691898.2001.11910537

Steiger, J. H. (1980). Tests for comparing elements of a correlation matrix. *Psychological Bulletin, 87*(2), 245–251. https://doi.org/10.1037/0033-2909.87.2.245

Stevens, S. S. (1946). On the theory of scales of measurement. *Science, 103*, 677–680. https://doi.org/10.1126/science.103.2684.677

Stigler, S. M. (1989). Francis Galton's account of the invention of correlation. *Statistical Science, 4*(2), 73–79. https://doi.org/10.1214/ss/1177012580

Tucker, L. R., Koopman, R. F., & Linn, R. L. (1969). Evaluation of factor analytic research procedures by means of simulated correlation matrices. *Psychometrika, 34*(4), 421–459. https://doi.org/10.1007/BF02290601

Tukey, J. W. (1965). The technical tools of statistics. *The American Statistician, 19*(2), 23–28. https://doi.org/10.1080/00031305.1965.10479711

Tukey, J. W. (1977). *Exploratory data analysis.* Addison-Wesley.

Waller, N. G. (2016). Fungible correlation matrices: A method for generating nonsingular, singular, and improper correlation matrices for Monte Carlo research. *Multivariate Behavioral Research, 51*(4), 554–568. https://doi.org/10.1080/00273171.2016.1178566

Wegman, E. J. (1990). Hyperdimensional data analysis using parallel coordinates. *Journal of the American Statistical Association, 85*(411), 664–675. https://doi.org/10.1080/01621459.1990.10474926

Wilkinson, L. (1999). *The grammar of graphics.* Springer. https://doi.org/10.1007/978-1-4757-3100-2

Wolf, F. M. (1986). *Meta-analysis: Quantitative methods for research synthesis* (Quantitative Applications in the Social Sciences No. 59). Sage.

Wright, S. (1934). The method of path coefficients. *Annals of Mathematical Statistics, 5*(3), 161–215. https://doi.org/10.1214/aoms/1177732676

INDEX